Analysis and Control of Underactuated Mechanical Systems

Amal Choukchou-Braham · Brahim Cherki ·
Mohamed Djemaï · Krishna Busawon

Analysis and Control of Underactuated Mechanical Systems

 Springer

Amal Choukchou-Braham
University of Aboubekr Belkaid
Chetouane, Tlemcen, Algeria

Brahim Cherki
University of Aboubekr Belkaid
Chetouane, Tlemcen, Algeria

Mohamed Djemaï
LAMIH ASHM
University of Valenciennes
  and Hainaut-Cambrésis
Valenciennes, France

Krishna Busawon
Faculty of Engineering and Environment
Northumbria University
Newcastle-upon-Tyne, UK

ISBN 978-3-319-37755-1   ISBN 978-3-319-02636-7 (eBook)
DOI 10.1007/978-3-319-02636-7
Springer Cham Heidelberg New York Dordrecht London

*To my beloved parents Amina Lokbani and
Abdelkader El Hathout*

Amal Choukchou-Braham

To my family

Brahim Cherki

*To my mother, my late father,
To my wife Badra, my children: Abdelaziz,
Yacine and Ayman,
To all my family.*

Mohamed Djemaï

To Christopher and Jennifer

Krishna Busawon

Preface

An underactuated mechanical system (UMS) is a system that has fewer control inputs than degrees of freedom. In contrast, a fully actuated mechanical system is one that has the same number of actuators as degrees of freedom. Underactuated mechanical systems arise in many real-life applications such as aircrafts, helicopters, spacecrafts, vertical take-off and landing aircrafts, underwater vehicles, mobile robots, walking robots, just to mention a few. Unlike fully actuated mechanical systems, the control of UMSs is quite a challenging task because the latter present a restriction on the control authority that makes the control design for these systems rather complicated. Also, very often it gives rise to complex theoretical problems that are not found in fully actuated systems and that cannot be solved using classical control techniques. In effect, some established results and properties of nonlinear systems such as feedback linearizability and passivity are no longer valid in the case of UMSs. Other undesirable properties like possessing an undetermined relative degree or being in a non-minimum phase are also customarily present in these systems. Moreover, several of these systems present a structural obstruction to the existence of smooth time invariant stabilizing control laws. Also, it is generally not easy to determine the controllability, at least locally, for these systems and even when they are controllable, the control laws can be discontinuous, periodic, and variant in time.

The control of UMSs has been investigated for quite a long time in the control literature and has been attracting more attention in recent years because of the growing interests in new robotic applications such as unmanned underactuated aerial or underwater vehicles. Different control strategies have been proposed in the literature, including backstepping, sliding mode, intelligent control, and much more. Several authors have attempted to present a classification and a generalization of these systems with the aim of proposing a systematic control design method for UMSs. Despite the diversity and large amount of research on the topic, it is difficult to highlight the structural properties of UMSs in a sufficiently general and exploitable manner that allows an unified treatment for the latter. As a matter of fact, there is no general theory that allows to conduct a systematic analysis and synthesis of control design for all UMSs. This has been the main motivating factor for us to write this current monograph.

The book presents theoretical explorations on the fundamental classification methods that are available in the literature; namely, the control flow diagram (CFD)-based classification of Seto and Baillieul and the structural properties-based classification of Olfati-Saber. It also proposes some tools for the systematic control design for underactuated systems. It aims to present a reference material for researchers and students working in the field of underactuated mechanical control. As such, the book is primarily intended for researchers and engineers in the system and control community. It can also serve as a complementary reading for post-graduated students studying control system theory.

Tlemcen, Algeria
August 2013

Amal Choukchou-Braham
Brahim Cherki
Mohamed Djemaï
Krishna Busawon

Contents

Biographies

Amal Choukchou-Braham received her engineering degree, M.Sc. and Ph.D. degree from University of Tlemcen, Algeria, in 1996, 1998, and 2011, respectively. Since 1999, she has held teaching and research positions in the Department of Electrical Engineering. She is a member of Tlemcen Automatic Laboratory (LAT) in the control team. Her research interests are in linear and nonlinear control and mechanical systems.

Brahim Cherki is currently professor of automatic control at Tlemcen University, Algeria. He received an engineer degree from Ecole Supérieure d'Electricité (Paris), a magister degree from Tlemcen University both in automatic control and a Ph.D. degree in robotics from Ecole Centrale de Nantes (France) in 1996. His main research areas are linear and nonlinear control and he is currently involved in projects on the control of waste water plants.

Mohamed Djemaï is currently full Professor at the University of Valenciennes, France, since 2008. He is with LAMIH-Laboratory of Industrial and Human Automation, Mechanics and Computer Science, CNRS (National Center for Scientific Research), UMR-8201. He received his Ph.D. degree in Control Systems Engineering from Paris XI University (France) in 1996. He was assistant professor with University of Versailles and CNAM (Conservatoire National des Arts et Metiers), Paris. He then joined the ENSEA (Ecole National Supérieure de l'Electronique et ses Applications) at Cergy Pontoise, France, where he took the role of Associate Professor from 2000 until 2008.

Prof. Djemaï research interests are in nonlinear control systems theory in general, with principal interest in variable structure systems, and sliding mode approach: in control, observation, fault detection, and the analysis and modeling of hybrid dynamical systems, which is the topic of the present research. He has applied his research to various industrial applications including power systems, chaotic and communication systems, biomechanical systems, robotics, and vehicle. He is member of IFAC TC 2.1 control system, IFAC TC. 1.3 on Discrete Event and Hybrid Systems. He is also IEEE senior Member. Prof. Djemaï is actually visiting professor at Northumbria University (2009–2013).

He has published more than 120 papers in his area of research. He was the co-editor of IFAC'CHAOS'06 Conference, co-authored two books and two special issues. He has supervised 12 successful Ph.D. students and is currently supervising four students. He has also secured several grants and participates in National and European Projects.

He has recently organized the first International Symposium of Environment Friendly Energies in Electrical applications (EFEEA 2010, www.univ-valenciennes.fr/efeea2010), and he was Co-General chair of EFEA 2012, in Newcastle (http://soe.northumbria.ac.uk/efea2012) and he is a Co-General chair of IEEE-ISCS 2013, 2nd International Conference on Systems and Control, Algiers, Algeria, 29–31 October 2013 (http://lias.labo.univ-poitiers.fr/icsc2013/). He was member for IPC and Organizing Committee in several international conferences.

Krishna Busawon is a Professor in Control systems Engineering and he is currently the head of Nonlinear Control research in the Faculty of Engineering and Environment. His research interest lies mainly in the area of mathematical modeling, nonlinear control and observer design, fault detection and isolation with application to various engineering disciplines such as mechanical, power and biotechnological systems. His recent research interests are in compressive sensing, chaos communications and hybrid systems. He obtained his first degree in Mathematics and Fundamental Sciences from University of St-Etienne in 1989. He then went to University of Lyon where he obtained his B.Eng. and M.Sc. Degree in Electrical Engineering in 1990 and 1991, respectively. He continued his post-graduate studies in the same university and consequently he obtained his M.Phil. and Ph.D. degree in Control Systems Engineering in 1992 and 1996, respectively. After his Ph.D. he was appointed as a Research Fellow at Simon Fraser University in 1997. He then joined the University of Nuevo Leon Mexico where he worked as a Lecturer in the Department of Mechanical and Electrical Engineering (FIME). In the year 2000, he joined Northumbria University where he was appointed as a Senior Lecturer in the School of Computing, Engineering and Sciences. Later, in 2006, he became a Reader in Control Systems Engineering and recently a Professor since February 2013 at the same university. He is currently the principal investigator of five Ph.D. students.

Acronyms

BIBS   Bounded Input Bounded State
CFD   Control Flow Diagram
DOF   Degrees of Freedom
GAS   Globally Asymptotically Stable
GES   Globally Exponentially Stable
ODE   Ordinary Differential Equation
SPD   Symmetric Positive Definite
UMS   Underactuated Mechanical System

Chapter 1
Introduction

…The highest education is that which does not merely give us information but makes our life in harmony with all existence.

Rabindranath Tagore

The control of underactuated mechanical systems is an active field of research in robotics and control system engineering. An underactuated mechanical system (UMS) is a system that has fewer actuators than configuration variables. Many real-life mechanical systems, including aircrafts, helicopters, spacecrafts, vertical take-off and landing aircrafts, underwater vehicles, mobile robots, walking robots, flexible systems, and nonlinear control benchmarks are examples of underactuated systems. The origin of underactuation is multiple: it can be natural due to the dynamics of the systems under study, or it can be artificial either by design or by deliberately removing actuators for the purpose of building challenging systems control or finally due to actuators' failure. Very often the control of UMSs gives rise to complex theoretical problems that cannot be solved using classical control techniques. In effect, the restriction on the control authority makes the control design for these systems rather complicated. Some established results and properties of nonlinear systems such as feedback linearizability and passivity are no longer valid in the case of UMSs. Other undesirable properties like possessing an undetermined relative degree or being in a non-minimum phase are also customarily present in these systems. Moreover, several of these systems present a structural obstruction to the existence of smooth time invariant stabilizing control laws. In effect, it is generally not easy to determine the controllability, at least locally, for these systems and even when they are controllable, the control laws can be discontinuous, periodic and time varying.

Even though these difficulties suggest that the control design for UMSs is challenging, the very existence of these systems together with their corresponding applications attracted the attention of many researchers, thereby compelling them to investigate the subject matter rigorously. As a matter of fact, the recent interest in new robotic applications involving unmanned underactuated vehicles, such as unmanned aerial or underwater vehicles, has provided a strong incentive and motivation to further develop research in this field. At the same time, this has also enabled researchers to tackle some underlying complex theoretical problems related to UMSs. Control

A. Choukchou-Braham et al., *Analysis and Control of Underactuated Mechanical Systems*, DOI 10.1007/978-3-319-02636-7_1,

algorithms for underactuated systems can be considered as soft solutions for actuators failure in fully actuated systems that avoid redundancy. Hence, it can be applied in many applications where safety is critical and can contribute to the success of delicate missions. Such control procedure naturally brings some benefits in terms of weight and cost reduction, and hence may encourage manufacturers to directly design underactuated algorithms.

As UMSs present challenges that are not found in fully actuated systems, different control strategies have been proposed in the literature, including backstepping and forwarding control as investigated in [2, 4–7, 12, 13, 22, 23, 29], energy and passivity-based control as in [10, 15, 20, 24], sliding mode control in [1, 9, 18, 26, 28], hybrid and switched control in [3, 8, 11, 17, 21, 30], and intelligent and fuzzy control as in [14, 16, 27], just to mention a few.

Based on the diversity and large amount of research on the topic, it is difficult to highlight the structural properties of UMSs in a sufficiently general and exploitable manner that allows an unified treatment for the latter. In effect, there is no general theory that allows to conduct a systematic analysis and synthesis of control design for all UMSs. As a result, most of the time these systems have to be dealt with on a case by case basis.

The first generalization for underactuated systems analysis is due to Spong in [25], where it was proven that these systems can be partially linearized by feedback, at least locally. Spong proposed changes in the input function according to actuation variables that transform nonlinear models into partially linear ones, including actuated and unactuated subsystems. However, the new control appears in both transformed subsystems.

Next, *Seto and Baillieul* [23], gave a first classification for UMSs according to their corresponding control flow diagram (CFD), which reflects the way generalized forces are transmitted through degrees of freedom. For underactuated systems with two degrees of freedom, three structures are identified, namely: chain, tree, and isolated vertex. The respective names of these structures are associated to the serial connection of degrees of freedom with the control input in the first structure and the fact that the control is transmitted simultaneously to degrees of freedom in the second one, and to the fact that the control does not affect some degrees of freedom in the last one. For high-order systems, the number of structures increases to seven and are a combination of the three basic ones. The authors of this classification also defined the degree of complexity of control of underactuated systems in the context of CFD according to their position in a hierarchy ordering they established. From this ordering, it appears that systems with chain structures are less complicated than the other structures. Thus, *Seto and Baillieul* proposed a systematic backstepping control design procedure for that class of systems. Unfortunately, a systematic control design methodology for systems with tree or isolated vertices is still an open problem.

Finally, *Olfati-Saber* in his excellent work [19] gives a second classification for UMSs based on some system structural properties like kinetic symmetry, actuation mode, integrable normalized generalized momentums, and interacting inputs. According to these properties, the author proposed explicit coordinates changes to uncouple the subsystems into cascade nonlinear subsystems and linear subsystems with structural properties that are convenient for control design purposes. These are obtained from the application of explicit change of control due to Spong associated with partial linearization procedures. Thus, three principal normal forms are identified: strict feedback form, feedforward form, and nontriangular linear-quadratic form. The respective names of these forms are associated with the particular lower-triangular, upper-triangular, and nontriangular structure in which the state variables appear in the dynamics. A classification is then established on the basis of these different normal forms. In the case of underactuated systems with two degrees of freedom, three classes are defined, namely Class-I, Class-II, and Class-III associated with strict feedback, nontriangular quadratic, and feedforward forms, respectively. For high-order systems, the procedure leads to the definition of five others classes practically based on the same forms and on slightly modified forms. To bring these systems to those with low-order, the author applies a reduction process that transforms high-order systems into reduced nonlinear subsystem in cascade with a linear subsystem. For systems in strict feedback and feedforward normal forms, the control strategy proposed by *Olfati-Saber* consists in first stabilizing the reduced order system followed by a backstepping or a forwarding procedures. The control of the nontriangular normal forms is still a major open problem.

The purpose of the present book is to assemble the various research done in the area of classification and control for UMSs and to further contribute to this research by developing tools for the systematic control design for underactuated systems. It aims to present reference material for researchers and students working in the field of underactuated mechanical control. In effect, we propose to analyze and compare the two classifications in order to establish whether or not they are related in some way or other. Next, we address stabilization procedures of the open control problem of tree and isolated vertex structures in the first classification. The strategy employed is to merge the two classifications. Specifically, the systematic procedure of control established for the chain structure and the other structures using change of control and coordinates, deduced from the second classification, is extended.

In addition to this introduction, which has presented both the motivation and an overview of the control design problem of UMSs, the rest of the book is composed of four other chapters and four appendices.

- *Chapter 2: Generalities and State-of-the-Art on the Control of Underactuated Mechanical Systems*

  The second chapter gives a brief overview of the control of UMSs. It provides the definition of UMS, gives the various origins of underactuation, describes the problems generated by the lack of control inputs, and presents the motivation for the control of UMSs. Then, a brief state-of-the-art on various control design approaches applied to these systems is given.

- *Chapter 3: Underactuated Mechanical Systems from the Lagrangian Formalism*

  The third chapter gives a qualitative description of UMSs obtained from the Lagrangian formalism. The general model associated with these systems is presented. The concept of underactuation and the resulting control problems are explained. Systems with nonholonomic constraints are defined and the difference between nonholonomy and underactuation is highlighted. Next, a brief overview on various control approaches applied to UMSs is presented; in particular, those based on partial linearization. Finally, some typical examples of UMSs are presented at the end of the chapter.

- *Chapter 4*: *Classification of Underactuated Mechanical Systems*

  Chapter 4 is devoted to the study of the classifications of UMSs that currently exists in the literature; namely the classification according to *Seto–Baillieul* and *Olfati-Saber*. These classifications are done with the aim of finding common properties that can lead to a systematic control design procedure for UMSs. One of the objectives of this chapter is to discuss whether the two classifications are possibly related or not. The typical UMSs examples provided in Chap. 2 are classified according to two classifications and a comparative study is made.

- *Chapter 5*: *Control Schemes Design for Underactuated Mechanical Systems*

  Chapter 5 is divided into three parts: The first part is devoted to the presentation of the backstepping systematic algorithm for global asymptotic stabilization of UMSs with chain structure. The second part is concerned with the control design of UMSs with tree structure. Specifically, a systematic control design scheme for that class of system is given. Additionally, we have a subclass of UMSs with tree structure that can be transformed into chain structure, under some assumptions, is presented. A stabilization procedure is proposed for other subclasses of UMSs that do not satisfy these assumptions.

  In the last part of this chapter, some discussions and suggestions on control design for UMSs with isolated vertices structure are given. Systems with such structures are considered difficult to control since some states are not reachable. For each part, simulation results carried out with Mathlab-Simulink software corresponding to the application of the respective proposed algorithms are given to show their effectiveness.

To allow easier reading, some classical results often found in disparate references are introduced in the appendices.

In the first appendix, a theoretical background on nonlinear system stability and control is given where the main stability criteria of nonlinear systems are recalled and some nonlinear control techniques are briefly described. The second appendix gives a rapid presentation of limits of linearization and dangers of destabilization. In the third appendix, some classical definitions of differential geometry are recalled. Finally, some controllability concepts of dynamical systems are revisited in the last appendix.

References

1. L. Aguilar, I. Boiko, L. Fridman, R. Iriante, Output excitation via continuous sliding modes to generate periodic motion in underactuated systems, in *Proc. 45th IEEE Conf. on Decision and Control*, USA (2006), pp. 1629–1634
2. N.P. Aneke, Control of underactuated mechanical systems. Ph.D. thesis, Technische Universiteit Eidhoven, 2003

3. R.N. Banavar, V. Sankaranarayanan, *Switched Finite Time Control of a Class of Underactuated Systems* (Springer, Berlin, 2006)
4. A.M. Bloch, Stabilization of the pendulum on a rotor arm by the method of controlled Lagrangians, in *Proc. Int. Conf. Robotics and Automation* (1999), pp. 500–505
5. A.M. Bloch, N. Leonard, J.E. Marsden, Controlled Lagrangians and the stabilization of mechanical systems I: The first matching theorem. IEEE Trans. Autom. Control **45**(12), 2253–2270 (2000)
6. A. Choukchou-Braham, B. Cherki, A new control scheme for a class of underactuated systems. Mediterr. J. Meas. Control **7**(1), 204–210 (2011)
7. A. Choukchou-Braham, B. Cherki, M. Djemai, A backstepping procedure for a class of underactuated system with tree structure, in *Proc. IEEE Int. Conf. on Communications, Computing and Control Applications, CCCA'11*, Hammamet, Hypersciences (2011)
8. A. Choukchou-Braham, B. Cherki, M. Djemai, Stabilisation of a class of underactuated system with tree structure by backstepping approach. Nonlinear Dyn. Syst. Theory **12**(4), 345–364 (2012)
9. F. Fahimi, Sliding mode formation control for underactuated surface vessels. IEEE Trans. Robot. **23**(6), 617–6221 (2007)
10. I. Fantoni, R. Lozano, *Nonlinear Control for Underactuated Mechanical Systems* (Springer, Berlin, 2002)
11. R. Fierro, F.L. Lewis, A. Low, Hybrid control for a class of underactuated mechanical systems. IEEE Trans. Autom. Control **29**(6), 649–654 (1999)
12. J. Ghommam, Commande non linéaire et navigation des véhicules marins sous actionnés. Ph.D. thesis, Orléans University, France, 2008
13. F. Gronard, R. Sepulchre, G. Bastin, Slow control for global stabilization of feedforward systems with exponentially unstable Jacobian linearisation, in *Proc. 37th IEEE Conf. on Decision and Control*, USA (1998), pp. 1452–1457
14. T. Han, S. Sam Ge, T. Heng Lee, Uniform adaptive neural control for switched underactuated systems, in *IEEE Int. Symposium on Intelligent Control*, San Antonio (2008), pp. 1103–1108
15. G. Hu, C. Makkar, W.E. Dixon, Energy-based nonlinear control of underactuated Euler Lagrange systems subject to impacts. IEEE Trans. Autom. Control **52**(9), 1742–1748 (2007)
16. C. Lin, Robust adaptive critic control of nonlinear systems using fuzzy basis function network: an LMI approach. J. Inf. Sci., 4934–4946 (2007)
17. A.D. Mahindrakar, R.N. Banavar, A swinging up of the Acrobot based on a simple pendulum strategy. Int. J. Control **78**(6), 424–429 (2005)
18. S. Nazrulla, H.K. Khalil, A novel nonlinear output feedback control applied to the Tora benchmark system, in *Proc. 47th IEEE Conf. on Decision and Control*, Mexico (2008), pp. 3565–3570
19. R. Olfati-Saber, Nonlinear Control of Underactuated Mechanical Systems with Application to Robotics and Aerospace Vehicles. Ph.D. thesis, Massachusetts Institute of Technology, Department Electrical Engineering and Computer Science, 2001
20. R. Ortega, A. Loria, P. Nicklasson, H. Sira-Ramirez, *Passivity-Based Control of Euler Lagrange Systems* (Springer, Berlin, 1998)
21. M. Reyhanoglu, A. Bommer, Tracking control of an underactuated autonomous surface vessel using switched feedback. IEEE Ind. Electron. IECON, 3833–3838 (2006)
22. R. Sepulchre, M. Janković, P. Kokotović, *Constructive Nonlinear Control* (Springer, Berlin, 1997)
23. D. Seto, J. Baillieul, Control problems in super articulated mechanical systems. IEEE Trans. Autom. Control **39**(12), 2442–2453 (1994)
24. M. Spong, F. Bullo, Controlled symmetries and passive walking. IEEE Trans. Autom. Control **50**(7), 1025–1031 (2005)
25. M.W. Spong, Energy based control of a class of underactuated mechanical systems, in *IFAC World Congress* (1996), pp. 431–435
26. D.A. Voytsekhovsky, R.M. Hirschorn, Stabilization of single-input nonlinear systems using higher order compensating sliding mode control, in *Proc. 45th IEEE Conf. on Decision and Control and the European Control Conf.* (2005), pp. 566–571

27. R. Wai, M. Lee, Cascade direct adaptive fuzzy control design for a nonlinear two axis inverted pendulum servomechanism. IEEE Trans. Syst. **38**(2), 439–454 (2008)
28. R. Xu, U. Ozguner, Sliding mode control of a class of underactuated systems. Automatica **44**, 233–241 (2008)
29. C. Yemei, H. Zhengzhi, C. Xiushan, Improved forwarding control design method and its application. J. Syst. Eng. Electron. **17**(1), 168–171 (2006)
30. M. Zhang, T.J. Tarn, A hybrid switching control strategy for nonlinear and underactuated mechanical systems. IEEE Trans. Autom. Control **48**(10), 1777–1782 (2003)

Chapter 2
Generalities and State-of-the-Art on the Control of Underactuated Mechanical Systems

> *…I only wanted to expose, in this work, what I managed to do at this moment in time and which may be used as a starting point for other research of the same kind.*
>
> *M.A. Lyapunov*

Ever since time began, mankind has never stopped dreaming about traveling from one continent to another and about flying like a bird, exploring the depths of the ocean and conquering space. His ambitions have compelled him to search for and to realize and to even improve the means that will permit him to realize his objectives. Furthermore, it would be difficult or even impossible to achieve such objectives without having recourse to mechanical systems. Even though the research interest in mechanical systems goes far back to the time of Newton, Lagrange, Kepler, Hamilton, and many other famous researchers, actually this area of research is even more active due to its diverse applications in real life and in the industrial domain.

In fact, during the last few decades, a number of scientific, industrial, and military applications have instigated the analysis and the rigorous derivation of control algorithms for mechanical systems. This area of research has also attracted the attention of mathematicians since the majority of the systems possess a global nonlinear characteristic, and their linear approximation seems to be inadequate. In combining their efforts, the engineers and scientists have developed several control design methodologies that include linear control, optimal control, adaptive, and nonlinear control, and more recently robust control in order to take into account uncertainties in a practical and real life context. In fact, the interest in mechanical systems became even stronger when researchers realized that the latter can be underactuated.

2.1 Underactuated Mechanical Systems: Generalities and Motivations

A mechanical system is said to be underactuated when the number of control inputs is less than the number of degrees of freedom to be controlled. This class of systems has a varied and rich applications, at both the practical and the theoretical level, in various fields such as in robotics, aeronautical and spatial systems, marine and

A. Choukchou-Braham et al., *Analysis and Control of Underactuated Mechanical Systems*, DOI 10.1007/978-3-319-02636-7_2,
© Springer International Publishing Switzerland 2014

underwater systems, and flexible and mobile systems. In contrast to systems that have direct practical applications, the pendulum systems, the Acrobot, the Pendubot, the Tora and the ball and beam systems have a meaning in terms of benchmarks for nonlinear control where classical procedures cannot be applied.

The underactuation can be due to one of the following reasons [40]:

(i) It can be natural due to the dynamics of the systems such as those of aircrafts, helicopters, and underwater vehicles.
(ii) It can be imposed by design in order to reduce the costs and weight such as satellites with two thrusters and flexible-link robots.
(iii) It can be due to actuators' failure such as in aeroplanes and ships.
(vi) It can be artificially imposed in order to generate low-order complex nonlinear systems so as to gain insight on the control of high-order UMSs such as the inverted pendulum and all the above benchmark examples mentioned above.

The restriction of the control authority renders the control of these systems rather complicated. In some sense, the underactuation characteristics are even more difficult to handle than the nonlinear characteristics of the underlying system. As a matter of fact, some well-established results and properties for nonlinear systems such as linearization by feedback, passivity and matching condition are not generally valid in the case of UMSs. Furthermore, these systems show other undesirable properties like an undetermined relative degree or non-minimal phase behavior.

On the other hand, several UMSs present a structural obstruction to the existence of smooth and time-invariant stabilizing feedback control laws, since they do not satisfy the well-known and necessary condition of Brockett [11] for smooth time-invariant feedback stabilization, which is one of the most remarkable contributions in this area. Typically, a first indication of this obstruction comes from the fact that the linearization of these systems around any equilibrium point is uncontrollable, particularly in the absence of gravity terms. Hence, false conclusions on the controllability can be easily drawn.

Although these control difficulties suggest that the objective of asymptotic stabilization is, without any doubt, too demanding for the control of UMSs, the very existence of these systems and the theoretical challenges they present have forced many researchers to fully investigate that topic. In addition, mastering the control of these systems can transform their shortcomings into advantages. In effect, for the same configuration space, a fully actuated system requires more controls than if it were underactuated. This increases the weight and cost of the system. Finding the means to control a version of an underactuated system allows to eliminate certain control devices, improves global performances, and reduces the cost of realization.

Additionally, underactuation provides a control solution for the safety of systems. For example, if a fully actuated system becomes faulty and if we have an underactuated control system, then we can use the latter in critical situations (as for example in the case of a fault in one of the thrusters of an aeroplane, rocket or space engine) in order to avoid complete failure of the system or mission. Obviously, such a solution is more economical than the addition of redundant actuators.

On the other hand, UMSs has been studied on a case by case basis due to the difficulty in putting forward sufficiently general and exploitable structural properties in order to classify them according to their corresponding properties, and, at the same time, to be able to choose the appropriate control strategy according to their classification. Hence, there have been various research works on the control synthesis and strategies of control for these systems.

2.2   Brief State-of-the-Art on the UMSs Control

The aim of this section is not to give a complete account on the literature on the control of UMSs but to highlight the main contributions in this area.

Among the most recognized works, there are those based on the energy point of view. These are mainly the works of Astrom, Bloch, Furuta, Spong, and others [3, 5, 7, 8, 10, 17, 27, 29, 34, 60–62].

In these works the general control strategy is to swing the systems (mainly of pendular types such as the Acrobot, the Pendubot, inertial wheel pendulum) in order to bring them to the neighborhood of their linearity domain. Once this domain is attained, a switch towards a linear control of LQR type or pole placement is realized.

In a similar fashion, certain passivity-based methods also consist in swinging or steering the previous systems but this time in order to bring them to their homocline orbits. After that, a switch towards a linear control is realized such as in the works of Fantoni, Ortega and Spong in [18, 41, 43, 60]. Other work on passivity due to Janković and Sepulchre relates to the transformation of the systems in a cascaded form [31, 56] such as for the Tora system or for the Pendubot, as in the work of Kolesnichenko [32].

Most of the time, the authors do not deem it necessary to establish a stability proof of the system with switch. Additionally, the application domain of these methods are quite restrictive in real applications.

Because of its complexity, the ball and beam system has been the subject of several studies, namely by using: methods of approximate linearization by Hauser et al. [25], saturation for stabilization of cascaded system in feedforward by Teel [65], stabilization by output feedback of Teel and Praly [66], small gains synthesis by Sepulchre [55] and sliding mode control by Voytsekhovsky and Hirschorn [68].

The VTOL (vertical take-off and landing aircraft) is another example of UMS that is largely studied, namely for its industrial applications and for its non-minimum phase property [18, 26, 36] and [14, 39].

Due to their wide application in industry, cranes, and inertia wheel pendulums have been studied extensively. Reviews on models, applications, and control strategies are presented and discussed, respectively, in [1] and in [9].

Marine and underwater vehicles have also been the subject of numerous research. For instance, a smooth and continuous control allowing to exponentially reach a desired position and orientation has been introduced by Egeland [15]. A periodic control that asymptotically stabilizes the vehicle to the origin has been presented by Pettersen and Egeland [45]. In addition, inspired by the work of Morin and Samson [37], Pettersen and Egeland [46] have proposed a periodic and non-stationary control allowing to obtain an exponential stability of the underactuated marine vehicle. Then, Pettersen and Nijmeijer [47] have proposed a time-varying control law that led to a global and practical tracking and stabilization of the underactuated marine vehicle. The work of Ghommam [21] formulates and solves dynamic control positioning problems and trajectory tracking of underactuated marine vehicles.

In addition to the problem of stabilization of UMSs, the problem of trajectory tracking has also been tackled in the works of Bullo, Hu, Reyhanoglu and Sandoz, [35, 44, 48, 52, 72]. On the other hand, some researchers focused their attention to the case where the condition of Brockett (on the stabilizability of nonlinear systems using time-invariant continuously differentiable state feedbacks) is not satisfied and have proposed discontinuous control algorithms. Among these works, we can cite those of Oriolo and Nakamura and those of Reyhanoglu [42, 49, 50].

Other control strategies have also be derived such as: backstepping and forwarding procedures by Gronard, Sepulchre and Seto [23, 56, 57, 71]; sliding mode control by Fridman, Fahimi, Khalil and Su [2, 16, 38, 64, 68, 70]; hybrid and switching control by Fierro, Tomlin and Zhang [19, 48, 67, 73], optimization-based design by [53, 54, 63], inverse dynamics control and differential flatness by [4, 6, 20, 51, 58], and fuzzy logic and neural networks by Han, Lin and Wai [24, 33, 69].

Recently, some researchers have been interested in the control of biped robots. For this one can cite the work of Chevallereau [13], Chemori [12], and that of Spong [22, 28, 30, 59].

2.3 Scope and Objectives of This Book

One can clearly notice that all the previous aforementioned systems have been studied on a case by case basis. Based on that observation, the main objective of this book is to attempt to find and present the means that will permit the synthesis of control laws in a systematic manner for all UMSs but not necessarily with the same type of control. To meet this objective, it is quite intuitive to look for common (or even different) properties of UMSs that will permit to classify them.

This book also aims to gather existing classifications for UMSs in the literature. In fact, there exist two such classifications. The first classification is due to *Dambing Seto and John Baillieul* [57], which is of a graphical nature. It consists in tracing the Control Flow Diagram (CFD) of the given system and describes the ways the control inputs are transmitted through the degrees of freedom. According to this approach, three main structures are identified, namely: the chain structure, the tree structure, and the isolated vertex (or point) structure.

The combination of these structures yields seven structures for this classification. The authors of this classification have proposed a systematic control procedure of backstepping type that can globally and asymptotically stabilize the systems belonging to the chain structure. The stabilization problem for the other two classes are still open problems according to them.

The second classification is due to *Reza Olfati-Saber* [40] and is rather of an analytical nature. It considers structural properties of mechanical systems such as the actuation of certain degrees of freedom, the coupling between the inputs and the integrability of generalized momentums. Thus, eight classes are generated among which three are considered to be the principal ones, namely: the strict feedback normal form, feedforward normal form, and the non-triangular normal form.

The author of this classification has proposed a control design procedure in two steps for the first two normal forms: first to stabilize the reduced system and then to extend the stabilization to the global system by a backstepping or by a forwarding procedure depending on the considered normal form.

Some control design suggestions have been given for the third form. However, the procedure proposed for the stabilization of the reduced system requires the verification of a rather restrictive hypothesis.

This book tries to give some answers to the stabilization of the tree and isolated vertex structures based on the *Seto and Baillieul* classification. These two structures are more difficult to control but have the advantage (or shortcoming, depending on one's viewpoint) of representing the majority of UMSs.

References

1. E.A.A. Rahman, A.H. Nayfeh, Z.N. Masoud, Dynamics and control of cranes: a review. J. Vib. Control **9**, 863–908 (2003)
2. L. Aguilar, I. Boiko, L. Fridman, R. Iriante, Output excitation via continuous sliding modes to generate periodic motion in underactuated systems, in *Proc. 45th IEEE Conf. on Decision and Control*, USA (2006), pp. 1629–1634
3. K.J. Astrom, K. Furuta, Swinging up a pendulum by energy control. Automatica **36**, 287–295 (2000)
4. A.H. Bajodah, D.H. Hodges, Y.H. Chen, Inverse dynamics of servo-contraints based on the generalized inverse. Nonlinear Dyn. **39**, 179–196 (2005)
5. R.N. Banavar, V. Sankaranarayanan, *Switched Finite Time Control of a Class of Underactuated Systems* (Springer, Berlin, 2006)
6. W. Blajer, K. Dziewiecki, K. Kolodziejczyk, Z. Mazur, Inverse dynamics of underactuated mechanical systems: a simple case study and experimental verification. Commun. Nonlinear Sci. Numer. Simul. **16**(5), 2265–2272 (2011)
7. A.M. Bloch, Stabilization of the pendulum on a rotor arm by the method of controlled Lagrangians, in *Proc. Int. Conf. Robotics and Automation* (1999), pp. 500–505
8. A.M. Bloch, N. Leonard, J.E. Marsden, Controlled Lagrangians and the stabilization of mechanical systems I: The first matching theorem. IEEE Trans. Autom. Control **45**(12), 2253–2270 (2000)
9. D.J. Block, K.J. Astrom, M.W. Spong, *The Reaction Wheel Pendulum* (Morgan and Claypool, San Rafael, 2007)
10. S. Bortoff, M.W. Spong, Pseudolinearization of the Acrobot using spline functions, in *Proc. IEEE Conf. on Decision and Control*, Tuscan (1992), pp. 593–598
11. R.W. Brockett, *Asymptotic Stability and Feedback Stabilization* (Birkhäuser, Basel, 1983)
12. A. Chemori, Quelques contributions à la commande non linéaire des robots marcheurs bipèdes sous actionnés. Ph.D. thesis, National Polytechnic Institute of Grenoble France, 2005
13. C. Chevallereau, G. Bessonnet, G. Abba, Y. Aoustin, *Les Robots Marcheurs Bipèdes Modélisation, Conception, Sythèse de la Marche, Commande* (Hermes Lavoisier, Paris, 2007)
14. W.E. Dixon, A. Behal, D. Dawson, S. Nagarkatti, *Nonlinear Control of Engineering Systems* (Birkhäuser, Basel, 2003)

15. O. Egeland, M. Dalsmo, O. Sordalen, Feedback control of a nonholonomic underwater vehicle with constant desired configuration. Int. J. Robot. Res. **15**, 24–35 (1996)
16. F. Fahimi, Sliding mode formation control for underactuated surface vessels. IEEE Trans. Robot. **23**(6), 617–6221 (2007)
17. Y. Fang, W.E. Dixon, D.M. Dawson, E. Zergeroglu, Nonlinear coupling control laws for an underactuated overhead crane system. IEEE/ASME Trans. Mechatron. **8**(3), 418–423 (2003)
18. I. Fantoni, R. Lozano, *Nonlinear Control for Underactuated Mechanical Systems* (Springer, Berlin, 2002)
19. R. Fierro, F.L. Lewis, A. Low, Hybrid control for a class of underactuated mechanical systems. IEEE Trans. Autom. Control **29**(6), 649–654 (1999)
20. M. Fliess, J. Lévine, P. Martin, P. Rouchon, Flatness and defect of nonlinear systems: introductory theory and examples. Int. J. Control **61**, 1327–1361 (1995)
21. J. Ghommam, Commande non linéaire et navigation des véhicules marins sous actionnés. Ph.D. thesis, Orléans University, France, 2008
22. R.D. Gregg, M.W. Spong, Reduction-based control with application to three-dimensional bipedal walking robots, in *Proc. American Control Conf.*, USA (2008), pp. 880–887
23. F. Gronard, R. Sepulchre, G. Bastin, Slow control for global stabilization of feedforward systems with exponentially unstable Jacobian linearisation, in *Proc. 37th IEEE Conf. on Decision and Control*, USA (1998), pp. 1452–1457
24. T. Han, S. Sam Ge, T. Heng Lee, Uniform adaptive neural control for switched underactuated systems, in *IEEE Int. Symposium on Intelligent Control*, San Antonio (2008), pp. 1103–1108
25. J. Hauser, S. Sastry, P. Kokotović, Nonlinear control via approximate input-output linearization. IEEE Trans. Autom. Control **37**(3), 392–398 (1992)
26. J. Hauser, S. Sastry, G. Meyer, Nonlinear control design for slightly non-minimum phase systems. Automatica **28**, 665–679 (1992)
27. T. Henmi, M. Deng, A. Inoue, Swing-up control of the Acrobot using a new partial linearization controller based on the Lyapunov theorem, in *Proc. IEEE Int. Conf. on Networking, Sensing and Control* (2006), pp. 60–65
28. J.K. Holm, M.W. Spong, Kinetic energy shaping for gait regulation of underactuated bipeds, in *Proc. 17th IEEE Int. Conf. on Control Applications*, San Antonio (2008), pp. 1232–1238
29. G. Hu, C. Makkar, W.E. Dixon, Energy-based nonlinear control of underactuated Euler Lagrange systems subject to impacts. IEEE Trans. Autom. Control **52**(9), 1742–1748 (2007)
30. Y. Igarashi, T. Hatanaka, M. Fujita, M.W. Spong, Passivity-based output synchronization in SE(3), in *Proc. American Control Conf.*, USA (2008), pp. 723–728
31. M. Jankovic, D. Fontaine, P.V. Kokotović, Tora example: cascade and passivity-based control designs. IEEE Trans. Control Syst. Technol. **4**(3), 292–297 (1996)
32. O. Kolesnichenko, A.S. Shiriaev, Partial stabilization of underactuated Euler Lagrange systems via a class of feedback transformations. Syst. Control Lett. **45**(2), 121–132 (2002)
33. C. Lin, Robust adaptive critic control of nonlinear systems using fuzzy basis function network: an LMI approach. J. Inf. Sci., 4934–4946 (2007)
34. A.D. Mahindrakar, R.N. Banavar, A swinging up of the Acrobot based on a simple pendulum strategy. Int. J. Control **78**(6), 424–429 (2005)
35. S. Martinez, J. Cortés, F. Bullo, Analysis and design of oscillatory control systems. IEEE Trans. Autom. Control **48**(7), 1164–1177 (2003)
36. N.H. McClamroch, I. Kolmakovsky, A hybrid switched mode control for V/STOL flight control problems, in *Proc. IEEE Conf. on Decision and Control and the European Control Conf.*, Japan (1996), pp. 2648–2653
37. P. Morin, C. Samson, Time-varying exponential stabilization of the attitude of a rigid spacecraft with two controls, in *Proc. 34th IEEE Conf. on Decision and Control*, New Orleans (1995), pp. 3988–3993
38. S. Nazrulla, H.K. Khalil, A novel nonlinear output feedback control applied to the Tora benchmark system, in *Proc. 47th IEEE Conf. on Decision and Control*, Mexico (2008), pp. 3565–3570

39. M. Oishi, C. Tomlin, Switching in nonminimum phase systems: application to a VSTOL aircraft, in *Proc. American Control Conf.* (2000), pp. 487–491
40. R. Olfati-Saber, Nonlinear control of underactuated mechanical systems with application to robotics and aerospace vehicles. Ph.D. thesis, Massachusetts Institute of Technology, Department Electrical Engineering and Computer Science, 2001
41. J.P. Oliver, O.A. Ramirez, Control based on swing up and balancing scheme for an underactuated system, with gravity and friction compensator, in *IEEE Fourth Congress of Electronics, Robotics and Automotive Mechanics* (2007), pp. 603–607
42. G. Oriolo, Y. Nakamura, Control of mechanical systems with second order nonholonomic constraints: underactuated manipulators, in *Proc. IEEE Conf. on Decision and Control*, UK (1991), pp. 2398–2403
43. R. Ortega, A. Loria, P. Nicklasson, H. Sira-Ramirez, *Passivity-Based Control of Euler Lagrange Systems* (Springer, Berlin, 1998)
44. Y. Peng, J. Han, Q. Song, Tracking control of underactuated surface ships: using unscented Kalman filter to estimate the uncertain parameters, in *Proc. Int. Conf. on Mechatronics and Automation*, China (2007), pp. 1884–1889
45. K. Pettersen, O. Egeland, Position and attitude control of an underactuated autonomous underwater vehicle, in *Proc. IEEE Conf. on Decision and Control*, Japan (1995), pp. 987–991
46. K. Pettersen, O. Egeland, Robust control of an underactuated surface vessel with thruster dynamics, in *Proc. American Control Conf.*, New Mexico (1997), pp. 3411–3415
47. K. Pettersen, H. Nijmeijer, Tracking control of an underactuated surface vessel, in *Proc. 37th Conf. on Decision and Control*, USA (1998), pp. 4561–4566
48. M. Reyhanoglu, A. Bommer, Tracking control of an underactuated autonomous surface vessel using switched feedback. IEEE Ind. Electron. IECON, 3833–3838 (2006)
49. M. Reyhanoglu, A. Van der Schaft, N.H. McClamroch, I. Kolmanovsky, Dynamics and control of a class of underactuated mechanical systems. IEEE Trans. Autom. Control **44**(9), 1663–1671 (1999)
50. M. Reyhanoglu, S. Cho, N.H. McClamroch, I. Kolmanovsky, Discontinuous feedback control of a planar rigid body with an unactuated internal degree of freedom, in *Proc. IEEE Conf. on Decision and Control*, Florida (1998), pp. 433–438
51. P. Rouchon, Flatness based control of oscillators. Z. Angew. Math. Mech. **85**(6), 411–421 (2005)
52. E.E. Sandoz, P.V. Kokotović, J.P. Hespanha, Trackability filtering for underactuated systems, in *Proc. American Control Conf.*, USA (2008), pp. 1758–1763
53. R. Seifried, Optimization-based design of minimum phase underactuated multibody system. Comput. Methods Appl. Sci. **23**, 261–282 (2011)
54. R. Seifried, Integrated mechanical and control design of underactuated multibody systems. J. Appl. Nonlinear Dyn. **67**(2), 1539–1557 (2012)
55. R. Sepulchre, Slow peaking and low-gain designs for global stabilization of nonlinear systems. IEEE Trans. Autom. Control **45**(3), 453–461 (1999)
56. R. Sepulchre, M. Janković, P. Kokotović, *Constructive Nonlinear Control* (Springer, Berlin, 1997)
57. D. Seto, J. Baillieul, Control problems in super articulated mechanical systems. IEEE Trans. Autom. Control **39**(12), 2442–2453 (1994)
58. H.J. Sira Ramirez, S.K. Agrawal, *Differentially Flat Systems* (Marcel Dekker, New York, 2004)
59. M. Spong, F. Bullo, Controlled symmetries and passive walking. IEEE Trans. Autom. Control **50**(7), 1025–1031 (2005)
60. M.W. Spong, Energy based control of a class of underactuated mechanical systems, in *IFAC World Congress* (1996), pp. 431–435
61. M.W. Spong, D.J. Block, The Pendubot: a mechatronic system for control research and education, in *Proc. of the 34th IEEE Conf. on Decision and Control*, New Orleans (1995)
62. M.W. Spong, P. Corke, R. Lozano, Nonlinear control of the inertia wheel pendulum. Automatica **37**, 1845–1851 (1999)

63. W. Steiner, S. Reichl, The optimal control approach to dynamical inverse problems. J. Dyn. Syst. Meas. Control **134**(2), 021010 (2012)
64. C.Y. Su, Y. Stepanenko, Sliding mode control of nonholonomic systems: underactuated manipulator case, in *IFAC Nonlinear Control Systems*, Tahoe City (1995), pp. 609–613
65. A.R. Teel, Saturation to stabilize a class of single-input partially linear composite systems, in *IFAC NOLCOS'92 Symposium* (1992), pp. 369–374
66. A.R. Teel, L. Praly, Tools for semiglobal stabilization by partial state and output feedback. SIAM J. Control Optim. **33**(5), 1443–1488 (1995)
67. C.J. Tomlin, S.S. Sastry, Switching through singularities, in *Proc. 36th IEEE Conf. on Decision and Control* (1997), pp. 1–6
68. D.A. Voytsekhovsky, R.M. Hirschorn, Stabilization of single-input nonlinear systems using higher order compensating sliding mode control, in *Proc. 45th IEEE Conf. on Decision and Control and the European Control Conf.* (2005), pp. 566–571
69. R. Wai, M. Lee, Cascade direct adaptive fuzzy control design for a nonlinear two axis inverted pendulum servomechanism. IEEE Trans. Syst. **38**(2), 439–454 (2008)
70. R. Xu, U. Ozguner, Sliding mode control of a class of underactuated systems. Automatica **44**, 233–241 (2008)
71. C. Yemei, H. Zhengzhi, C. Xiushan, Improved forwarding control design method and its application. J. Syst. Eng. Electron. **17**(1), 168–171 (2006)
72. H. Yu, Y. Liu, T. Yang, Tracking control of a pendulum-driven cart-pole underactuated system, in *Proc. IEEE Int. Conf. on Systems, Man and Cybernetics*, Japan (2007), pp. 2425–2430
73. M. Zhang, T.J. Tarn, A hybrid switching control strategy for nonlinear and underactuated mechanical systems. IEEE Trans. Autom. Control **48**(10), 1777–1782 (2003)

Chapter 3
Underactuated Mechanical Systems from the Lagrangian Formalism

> *…Happy families are all alike, every unhappy family is unhappy in its own way.*
>
> *Léon Tolstoï, Anna Karenina*

When one is interested in controlling systems for which the nonlinear dynamics cannot be neglected, it has been well-known, since the time of Poincaré, that these nonlinear systems have extremely complex behaviors so that the application of a particular design method in control theory might not be suitable. It is therefore necessary to clarify, somehow, the class of systems we are interested in.

In this chapter, we are interested in the class of UMSs that are derived from the Lagrangian formalism. As a result, the first part of this chapter is dedicated to an introduction on Lagrangian systems. After that, the notion of underactuation is explained. Then, we give the definition of non-holonomic constraints as well as highlight the differences and subtleties that exist between underactuation and non-holonomy. We demonstrate why the control of UMSs leads to challenging theoretical problems, some of which are still open till now. Finally, the end of this chapter is dedicated to the presentation of the models of some UMSs.

3.1 Lagrangian System

A Lagrangian system is a system whose behavior is described by Euler–Lagrange's equations. They are defined by a set of nonlinear ordinary differential equations (ODEs). The Lagrangian formalism is a powerful mathematical modeling tool based on the variational method to model a large class of physical systems; in particular mechanical systems. A thorough review on variational modeling of mechanical and electro-mechanical systems can be found in the following respective references [11, 12] and [18, 37].

Underactuated systems being the basis of mechanical systems, their modeling can be done using Euler–Lagrange's equations. For a n degrees of freedom (DOF) system, the Euler–Lagrange equations are given by [23]

$$\frac{d}{dt}\frac{\partial L(q,\dot{q})}{\partial \dot{q}} - \frac{\partial L(q,\dot{q})}{\partial q} = F(q)u \tag{3.1}$$

A. Choukchou-Braham et al., *Analysis and Control of Underactuated Mechanical Systems*, DOI 10.1007/978-3-319-02636-7_3,
© Springer International Publishing Switzerland 2014

where $q \in Q$ denotes the configuration vector which belongs to an n-dimensional configuration manifold, $u \in R^m$ and $F(q)$ is the external forces matrix. $L(q, \dot{q})$ is the Lagrangian associated to the given system and expressed by the difference between its kinetic and potential energies:

$$L(q, \dot{q}) = K - V = \frac{1}{2} \dot{q}^T M(q) \dot{q} - V(q) \tag{3.2}$$

where K denotes the kinetic energy, V the potential energy, and $M(q)$ the positive definite inertia matrix. The notation $\dot{q}^T$ denotes the transpose of $\dot{q}$.

The advantage of the Lagrangian formalism is that the form of Euler–Lagrange's equations is invariant. Moreover, it allows to directly obtain the equation describing the evolution of mechanical systems as a function of the applied forces.

Based on these equations, the equations of motion of a mechanical system can be deduced as follows:

$$\sum_j m_{kj}(q) \ddot{q}_j + \sum_{i,j} \Gamma_{ij}^k(q) \dot{q}_i \dot{q}_j + g_k(q) = e_k^T F(q) u, \quad k = 1, \ldots, n$$

where e_k is the kth standard basis in R^n, $g_k(q) = \partial_{q_k} V(q)$, and $\Gamma_{ij}^k(q)$ the Christoffel symbol defined by

$$\Gamma_{ij}^k(q) = \frac{1}{2} \left(\frac{\partial m_{kj}(q)}{\partial q_i} + \frac{\partial m_{ki}(q)}{\partial q_j} - \frac{\partial m_{ij}(q)}{\partial q_k} \right)$$

Under a vectorial form, we can write

$$M(q) \ddot{q} + C(q, \dot{q}) \dot{q} + G(q) = F(q) u \tag{3.3}$$

$c_{ij} = \sum_{k=1}^{n} \Gamma_{kj}^i(q) \dot{q}_k$ is an element of $C(q, \dot{q})$. The term $C(q, \dot{q}) \dot{q}$ regroups two terms involving $\dot{q}_i \dot{q}_j$ named centrifugal ($i = j$) and Coriolis ($i \neq j$), and $G(q)$ contains the gravity terms (for more details see [33]).

The matrix defined by $S_0 = \dot{M}(q) - 2C(q, \dot{q})$ is antisymmetric, which then allows to have $\dot{M}(q) = C(q, \dot{q}) + C^T(q, \dot{q})$. Taking this property into account and the fact that $M(q)$ is a symmetric positive definite (SPD) matrix, one can introduce the *Legendre transform* defined by

$$p = \frac{\partial L}{\partial \dot{q}} = M(q) \dot{q}$$

In this case, the dynamics (3.3) can be rewritten under the following canonical form:

$$\begin{cases} \dot{q} = M^{-1}(q) p \\ \dot{p} = -G(q) + \bar{C}^T(q, p) M^{-1}(q) p + F(q) u \end{cases} \tag{3.4}$$

where $\bar{C}^T(q, p) = C(q, M^{-1} p)$.

Equation (3.4) is called *Legendre normal form* for a mechanical system. Setting $x_1 = q$ and $x_2 = p$, (3.4) can be rewritten as

$$\begin{cases} \dot{x}_1 = M^{-1}(x_1)x_2 \\ \dot{x}_2 = -G(x_1) + x_2^T Q(x_1)x_2 + F(x_1)u \end{cases} \tag{3.5}$$

or in compact form as

$$\dot{x} = f(x) + g(x)u \tag{3.6}$$

Remark 3.1 For mechanical systems Eqs. (3.3) and (3.4) are equivalent. However, the *Legendre normal form* is a first-order ODE whereas (3.6) is of second order.

Additionally, a mechanical system written in the form (3.6) belongs to the control-affine class of systems for which analytical methods for the analysis of controllability, observability, and the design control laws are available.

3.2 Fully Actuated Mechanical Systems

A mechanical system described by (3.1) is said to be fully actuated if $m = n$, which is equivalent to saying that $F(q)$ is invertible. For a fully actuated system, the number of inputs is equal to the dimension of their configuration space.

Consequently, these systems are linearizable via feedback and do not possess a zero dynamic. This can be demonstrated by applying the following control law:

$$u = F(q)^{-1}\big(M(q)v + C(q,\dot{q})\dot{q} + G(q)\big)$$

Setting $x_1 = q$ and $x_2 = \dot{q}$ we get

$$\dot{x}_1 = x_2$$

$$\dot{x}_2 = v$$

which is a double integrator system. This is why most of the fully actuated mechanical problems can be reduced to linear systems problems [23].

3.3 Underactuated Mechanical Systems

A controlled mechanical system with configuration vector $q \in Q$ and Lagrangian $L(q,\dot{q})$ satisfying Euler–Lagrange's equations (3.1) is called an underactuated mechanical system if $m < n$. In other words, UMSs are mechanical systems that have fewer independent inputs than the number of DOF to be controlled. Consequently, the generalized inputs cannot control the instantaneous acceleration in every directions [12, 36].

Fig. 3.1 Unicycle type
system

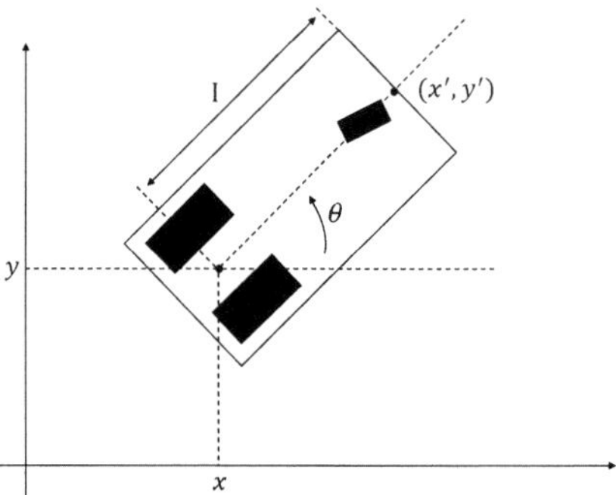

As mentioned in the previous chapter, underactuations may occur in different
fashions, the most evident one is due to the system dynamics itself. Typical examples
of these include: planes, helicopters, submarines, and locomotion systems without
wheels. Underactuations can also be due to design with the aim to reduce the load for
certain practical applications such as the satellites or flexible robots. The underactuation can also be caused by the failure of an actuator or can be artificially imposed
in order to generate complex systems of a fairly low order. This is the case for the
well-known inverted pendulum, the Acrobot, the Pendubot, the Tora, the ball and
beam and many more. All these systems will be presented at the end of this chapter.

For some UMSs, the lack of actuators is often interpreted as constraints on the
acceleration; that is, as second-order non-holonomic constraints.

3.4 Non-holonomic Mechanical Systems

A first-order mechanical system with non-holonomic[1] constraints is a Lagrangian
system having m velocity constraints $(m < n)$

$$W^T(q)\dot{q} = 0$$

non-integrable (where W is an $(m \times n)$ matrix), that is, there is no function $\varphi(t)$
such that $\dot{\varphi} = W^T(q)\dot{q}$.

These systems are characterized by the existence of non-integrable kinematic
constraints. Systems of unicycle types [1] Fig. 3.1, such as wheeled mobile robot,
wheeled vehicles or trailers vehicles are the most common examples.

According to [22, 38] when a mechanical system is subjected to first-order non-
holonomic constraints (velocity constraints), its dynamics can be written as

$$M(q)\ddot{q} + N(q, \dot{q}) = W(q)\lambda + F(q)u$$

$$W^T(q)\dot{q} = 0$$

---

[1] Holonomic: Greek word that signifies whole, integer.

where $\lambda \in R^m$ is the vector of Lagrange multipliers, the term $W(q)\lambda$ can be considered as the required force to maintain the constraints. The literature on non-holonomic systems is extremely vast, the reader can refer to [15] for an excellent review on this domain. One can also refer to [4, 36] for a quick review of the key concepts for the control of non-holonomic systems and other problems related to their kinematics.

Unlike non-holonomic systems with first-order constraints (velocity constraints) which are largely treated in the literature, UMSs are usually seen by many researchers [5, 10, 12, 23, 24, 26] as non-holonomic systems with second-order constraints or acceleration constraints.

According to these authors, for an UMS where the configuration variables can be partitioned, without loss of generality, into $q = (q_1, q_2)$, $q_1 \in \mathbb{R}^m$, $q_2 \in \mathbb{R}^{n-m}$ and if $F(q) = (I_m, 0)^T$, the Euler–Lagrange equations are given by

$$\begin{aligned}
m_{11}(q)\ddot{q} + m_{12}(q)\ddot{q} + N_1(q, \dot{q}) &= F(q)u \\
m_{21}(q)\ddot{q} + m_{22}(q)\ddot{q} + N_2(q, \dot{q}) &= 0
\end{aligned} \tag{3.7}$$

where $N_i(q, \dot{q})$ contains the centrifugal, Coriolis, and gravitational terms.

The second equation (3.7) represents the underactuated part of the system under the form of second-order constraints, generally non-integrable. In this case the constraints are not located at the kinematic level but at the dynamic level; since the number of independent actuators is less than the number of DOF. Common examples of these systems include marine and underwater vehicles, space robots, and underactuated articulated arms.

In contrast to the above, a mechanical system with holonomic constraints is one containing constraints which depend on the generalized coordinates (configuration variables) and time only.

Oriolo and Nakamura [24] gave the necessary and sufficient conditions for an UMS to contain non-holonomic constraints of second order or of first order or simply holonomic constraints.

3.5 Underactuation and Non-holonomy

The control of non-holonomic vehicles and that of underactuated vehicles have been the subject of distinct studies. This is partly justified by the difference in structures of the corresponding models. For non-holonomic systems, the difficulty (from the control engineering point of view) lies at the kinematic model level, while in the case of underactuated systems, the difficulty is rather related to its dynamics. This distinction equally implies a hierarchy as far as the difficulty in synthesizing control systems is concerned. While fairly general methods have been proposed for the control of non-holonomic systems (and more generally for the control of nonlinear systems), the control of underactuated systems is performed on a case by case basis due to the difficulty in finding sufficiently general and exploitable structural properties for the control synthesis.

The general relation between non-holonomic systems and underactuated systems is not completely established. Having said that, these two classes of systems possess numerous common points that are rarely explained in the literature, and the comprehension of which can permit progress towards a unified treatment of control problems associated with these systems.

The reader can find in the work of Jarzebowska [14] an excellent comparative and classification study of non-holonomic constraints of first order, qualified by the author as material constraints while non-holonomic second-order constraints due to underactuation are qualified as non-material constraints.

Remark 3.2

- When an UMS contains first-order non-holonomic constraints, a common practice for the control of these systems is to transform them into a canonical form that simplifies control design. This canonical form is in the so-called chain form (3.8) [15, 20, 21].

$$\dot{\xi}_1 = u_1$$
$$\dot{\xi}_2 = u_2 \tag{3.8}$$
$$\dot{\xi}_3 = \xi_2 u_1$$

Some systems that can be written in this form are mobile robots and traction vehicles. On the other hand, for UMSs that contain second-order non-holonomic constraints, some researchers have proposed to transform these systems into a second-order chain form (3.9) [1]

$$\ddot{\xi}_1 = u_1$$
$$\ddot{\xi}_2 = u_2 \tag{3.9}$$
$$\ddot{\xi}_3 = \xi_2 u_1$$

While first-order chained systems can have a dimension greater than that in (3.8), second-order chained systems do not exceed a dimension higher than 3. In addition, the latter are known to be more difficult to control than the first one.

- Some control engineers prefer to use the term underactuation constraints of second order rather than non-holonomic constraints of second order arguing the fact that, when an UMS is augmented by the missing actuators, it becomes a fully actuated system. It can then operate correctly while it is subjected to non-holonomic constraints, the fact of adding actuators cannot solve, for instance, a displacement problem in a certain direction. For example, to get rid of these constraints in a car would mean adding additional wheels, and this will bring a change in the kinematics of the system, which implies a change in the initial model.

- The general definition of UMS possesses some limitations. As a matter of fact, suppose that an UMS possesses a first-order non-holonomic constraint; for example consider the mobile robot of Fig. 3.1 with generalized coordinates (x, y, θ) represented by the following equations:

$$\dot{x} = v\cos\theta$$
$$\dot{y} = v\sin\theta \tag{3.10}$$
$$\dot{\theta} = u$$

the inputs are v: displacement velocity and u: angular velocity. The system is subjected to a non-integrable velocity constraint given by

$$\dot{x}\sin\theta - \dot{y}v\cos\theta = 0$$

This system with three DOF and two inputs is in fact an UMS. However, as it possesses a velocity constraint preventing a lateral displacement, it therefore follows that it is only possible to control two DOF while still having two inputs. Hence, in some sense, it is as if the system is fully actuated.

Whatever the cause and the consequence of underactuation, the control of the latter is extremely important due to their wide range of applications. However, the restriction on the number of control inputs gives rise to principal difficulties in the control design.

3.6 Problematics Associated with UMSs

Underactuated mechanical systems constitute a rich class of systems both in terms of applications and control problems. Effectively, these systems have attracted the attention of both control engineers and mathematicians because of their nonlinear characteristics and because of the problems related to underactuation. Unlike fully actuated systems whereby a number of results have been developed and applicable to some classes of systems—such as feedback linearization and passivity—very few results are valid for entire classes of UMSs since one or several of the aforementioned properties are no longer valid. In fact, it was shown [31] that these systems are not fully feedback linearizable, they are not completely decouplable, and they do not satisfy the passivity property and sometimes the matching condition. Classical control techniques such as computed torque, backstepping, passivity-based control, sliding mode control cannot be directly applied except for some specific cases. In addition, other properties such as an undetermined relative degree or a non-minimum phase behavior are apparent.

On the other hand, several UMSs present a structural obstruction to the existence of smooth and time-invariant stabilizing control, since it does not satisfy the well-known and necessary condition for smooth time-invariant feedback stabilization of Brockett [8], which is one of the most remarkable contributions in this domain.

Typically, a first indication to this obstruction comes from the fact that a linearization of these systems around any equilibrium point is not controllable (especially in absence of gravity terms). Consequently, false conclusions on the controllability of the nonlinear system can be drawn.

In this case, some researchers propose to employ a discontinuous control to overcome this problem [3, 9, 17, 19, 25, 27, 28, 30].

On the other hand, in the presence of potential terms, a local and exponential stabilization by a regular feedback, continuous and invariant is evidently possible.

Remark 3.3 For linear systems, controllability implies stabilizability. This is not true for nonlinear systems. The theorem of Brockett [8] gives a necessary condition on the stability of nonlinear systems by a continuous control.

Theorem 3.1 [8] *Consider a system given by*

$$\dot{x} = f(x, u) \tag{3.11}$$

with $f(0, 0) = 0$ and $f(\cdot, \cdot)$ is definite continuous in a neighborhood of the origin. Necessary conditions for the existence of a continuous time-invariant control and that renders the origin asymptotically stable are that:

(i) *The linearized system around the origin is stabilizable.*
(ii) *There exists a neighborhood V of the origin such that for every $\zeta \in V$, there exists a control input $u_\zeta(\cdot)$ defined on $[0, \infty[$ such that this control input steers the solution of $\dot{x} = f(x, u)$ from $x = \zeta$ at $t = 0$ towards $x = 0$ at $t = \infty$.*
(iii) *The mapping $\gamma : A \times \mathbb{R}^m \to \mathbb{R}^n$ defined by $\gamma : (x, u) \mapsto f(x, u)$ must be surjective in a neighborhood of the origin.*

The first condition represents the rank condition of a linear system. Note that in the linear case, the rank condition is a necessary and sufficient condition for the controllability and the existence of a continuous and differentiable control law for linear systems: $\dot{x} = Ax + Bu$.

The second property represents the controllability property in the nonlinear case. This condition is not sufficient to determine a control law with a certain regularity. Hence the necessity to introduce condition (iii), which corresponds to the necessary condition of this theorem. In fact, the third condition implies that the mapping must be locally surjective while the image of the mapping $(x, u) \mapsto f(x, u)$, for x and u arbitrarily close to 0, must contain a neighborhood of the origin.

To clarify this situation, consider for example the following system [10]:

$$\dot{x} = u = \varepsilon_1$$

$$\dot{y} = v = \varepsilon_2$$

$$\dot{z} = yu - xv = \varepsilon_3$$

Does there exist a continuous control $(u, v) = (u(x, y, z), v(x, y, z))$ that renders the origin of the above system asymptotically stable?

The third condition of Brockett signifies that the system must contain a solution (x, y, z, u, v) for each ε_i $(i = 1, 2, 3)$ in a neighborhood of the origin. This is not the case here since the system does not have a solution for $\varepsilon_3 \neq 0$ and $\varepsilon_1 = 0, \varepsilon_2 = 0$.

Unlike the previous example, the following example satisfies the third condition:

$$\dot{x} = u = \varepsilon_1$$

$$\dot{y} = v = \varepsilon_2$$

$$\dot{z} = xy = \varepsilon_3$$

Hence, for this particular example, there exists a continuous control law that asymptotically stabilizes the system at the origin.

In summary, the UMSs are not completely linearizable by feedback. Consequently, they are not completely decouplable. The controllability of these systems is difficult to demonstrate, and even if they are controllable, it does not imply that they are controllable with smooth and continuous control laws. Additionally, the lack of common and general properties between them implies that these systems are studied on a case by case basis.

All these difficulties of control design suggest that the objective of asymptotic stabilization is without doubt too demanding for the control of UMSs. Nevertheless, an interesting property that is valid for all these systems is the possibility of a partial linearization by feedback. This property was discovered by Spong [31] and is a consequence of the positive definiteness of the inertia matrix. This linearization may be collocated or non-collocated depending on the actuation or nonactuation of the linearized variables.

3.7 Partial Linearization by Feedback

Consider again the UMS model given by

$$M(q)\ddot{q} + C(q,\dot{q})\dot{q} + G(q) = F(q)\tau$$

where $\tau \in \mathbb{R}^m$ is the control input and $F(q) \in \mathbb{R}^{n \times m}$ is the non-square matrix of external forces with $m < n$.

Suppose that $F(q) = [0, I_m]^T$, then the configuration vector can be partitioned into $q = (q_1, q_2) \in \mathbb{R}^{n-m} \times \mathbb{R}^m$, where q_1 represents the non-actuated configuration vector and q_2 the actuated configuration vector, respectively. As a result of this partition, the inertia matrix and the model of the UMS take the following form:

$$\begin{bmatrix} m_{11}(q) & m_{12}(q) \\ m_{21}(q) & m_{22}(q) \end{bmatrix} \begin{bmatrix} \ddot{q}_1 \\ \ddot{q}_2 \end{bmatrix} + \begin{bmatrix} N_1(q,\dot{q}) \\ N_2(q,\dot{q}) \end{bmatrix} = \begin{bmatrix} 0 \\ \tau \end{bmatrix} \tag{3.12}$$

Due to the lack of control input in the first equation of (3.12), it is not possible to completely linearize this system by a change of coordinates. However, it is possible to partially linearize this system so that the dynamics of q_2 is transformed into a double integrator.

3.7.1 Collocated Partial Linearization

When the actuated dynamics is q_2, the linearization procedure of this dynamics is called collocated partial feedback linearization. In an equivalent fashion, this linearization can be considered as an input–output linearization with respect to the output $y = q_2$. Spong has shown that all underactuated systems of the form (3.12) can be partially linearized by using a change in input.

Proposition 3.1 [23] *There exists a global and invertible control of the form $\tau = \alpha(q)u + \beta(q,\dot{q})$ that partially linearizes the dynamics of (3.12):*

$$\begin{aligned}
\dot{q}_1 &= p_1 \\
\dot{p}_1 &= f_0(q,p) + g_0(q)u \\
\dot{q}_2 &= p_2 \\
\dot{p}_2 &= u
\end{aligned} \tag{3.13}$$

where $\alpha(q)$ is a symmetric positive definite (SPD) $(m \times m)$ matrix, $g_0(q) = -m_{11}^{-1}(q)m_{12}(q)$

Proof From the first line of Eq. (3.12), we have

$$\ddot{q}_1 = -m_{11}^{-1}(q)N_1(q,\dot{q}) - m_{11}^{-1}(q)m_{12}(q)\ddot{q}_2$$

which yields the expression of $g_0(q)$; note that m_{11} is invertible because of the positive definiteness of M.

By replacing this equation in the second line of (3.12), we obtain

$$(m_{22}(q) - m_{21}(q)m_{11}^{-1}(q)m_{12}(q))\ddot{q}_2 + N_2(q,\dot{q}) - m_{21}(q)m_{11}^{-1}(q)N_1(q,\dot{q}) = \tau$$

The proof is established by defining

$$\alpha(q) = \left(m_{22}(q) - m_{21}(q)m_{11}^{-1}(q)m_{12}(q)\right)$$

$$\beta(q,\dot{q}) = N_2(q,\dot{q}) - m_{21}(q)m_{11}^{-1}(q)N_1(q,\dot{q})$$

and by observing that $\alpha(q)$ is SPD. $\qquad\square$

3.7.2 Non-collocated Partial Linearization

The partial linearization procedure that linearizes the dynamics of the non-actuated configuration is called non-collocated partial feedback linearization. This partial linearization is possible if the number of control inputs is greater than or equal to the number of non-actuated configuration variables of the non-actuated configuration.

Consider the following non-actuated system:

$$\begin{bmatrix} m_{00}(q) & m_{01}(q) & m_{02}(q) \\ m_{10}(q) & m_{11}(q) & m_{12}(q) \\ m_{20}(q) & m_{21}(q) & m_{22}(q) \end{bmatrix} \begin{bmatrix} \ddot{q}_0 \\ \ddot{q}_1 \\ \ddot{q}_2 \end{bmatrix} + \begin{bmatrix} N_0(q,\dot{q}) \\ N_1(q,\dot{q}) \\ N_2(q,\dot{q}) \end{bmatrix} = \begin{bmatrix} \tau_0 \\ \tau_1 \\ 0 \end{bmatrix} \tag{3.14}$$

where $q = (q_0, q_1, q_2) \in \mathbb{R}^{n_0} \times \mathbb{R}^{n_1} \times \mathbb{R}^{n_2}$ with $n_1 = n_2 = m$ and $n_0 = n - 2m \geq 0$.

Proposition 3.2 [23] *Consider the underactuated system (3.14). There exists a change in input of the form $\tau = \alpha_1(q)u + \beta_1(q,\dot{q})$ that partially linearizes the dynamics of (3.14) in the set $U = \{q \in \mathbb{R}^n / \det(m_{21}(q) \neq 0)\}$*

$$\dot{q}_0 = p_0$$
$$\dot{p}_0 = u_0$$
$$\dot{q}_1 = p_1$$
$$\dot{p}_1 = f_0(q,p) + g_0(q)u_0 + g_2(q)u_2$$
$$\dot{q}_2 = p_2$$
$$\dot{p}_2 = u_2$$

where $\tau = (\tau_0, \tau_1)$, $u = (u_0, u_1)$ and $u_1 = \alpha_0(q)u_0 + \alpha_2(q)u_2 + \beta_2(q,\dot{q})$ with

$$f_0(q,p) = -m_{21}^{-1}(q)N_2(q,\dot{q})$$
$$g_0(q) = -m_{21}^{-1}(q)m_{20}(q)$$
$$g_2(q) = -m_{21}^{-1}(q)m_{22}(q)$$

The proof is based on that of collocated partial linearization. For more details, see [23].

3.7.3 Partial Linearization Under Coupled Inputs

Consider for the underactuated system (3.1) that $F(q)$ can be written as

$$F(q) = \begin{bmatrix} F_1(q) \\ F_2(q) \end{bmatrix}$$

such that $F_2(q)$ is a $(m \times m)$ invertible matrix and q can be decomposed into $(q_1, q_2) \in \mathbb{R}^{(n-m)} \times \mathbb{R}^m$.

The definition of coupled inputs implies that $F_1(q) \neq 0$ for all q.

Proposition 3.3 [23] *Consider the underactuated system with coupled inputs, that is* $(F_1(q) \neq 0, \det(F_2(q)) \neq 0$ *for all* $q)$,

$$\begin{bmatrix} m_{11}(q) & m_{12}(q) \\ m_{21}(q) & m_{22}(q) \end{bmatrix} \begin{bmatrix} \ddot{q}_1 \\ \ddot{q}_2 \end{bmatrix} + \begin{bmatrix} N_1(q,\dot{q}) \\ N_2(q,\dot{q}) \end{bmatrix} = \begin{bmatrix} F_1(q) \\ F_2(q) \end{bmatrix} \tau \tag{3.15}$$

and suppose that the matrix $\Lambda(q) = F_2(q) - m_{21}(q)m_{11}^{-1}(q)F_1(q)$ *is invertible for all* q. *Then, there exists a change in inputs* $\tau = \alpha(q)u + \beta(q,\dot{q})$ *that partially linearizes* (3.15) *under the form*

$$\dot{q}_1 = p_1$$
$$\dot{p}_1 = f_0(q,p) + g_0(q)u$$
$$\dot{q}_2 = p_2$$
$$\dot{p}_2 = u$$

where

$$\alpha(q) = \Lambda^{-1}(q)\big[m_{22}(q) - m_{21}(q)m_{11}^{-1}(q)m_{12}(q)\big]$$
$$\beta(q,\dot{q}) = \Lambda^{-1}(q)\big[N_2(q,\dot{q}) - m_{21}(q)m_{11}^{-1}(q)N_1(q,\dot{q})\big]$$
$$f_0(q,p) = m_{11}^{-1}(q)\big[F_1(q)\beta(q,\dot{q}) - N_1(q,\dot{q})\big]$$
$$g_0(q) = m_{11}^{-1}(q)\big[F_1(q)\alpha(q) - m_{12}(q)\big]$$

Remark 3.4

- For more details of the proof of these two last propositions, refer to [23] and [31].
- The partial linearization procedure in the proposition (3.3) is particularly used for autonomous underactuated system with six degrees of freedom such as airplanes and helicopters.

Another interesting property that is present in several UMSs is that of symmetry. In what follows, we are going to define the different notions of symmetry in relation to Lagrangian systems.

3.8 Symmetry in Mechanics

A Lagrangian $L(q,\dot{q})$ is symmetric with respect to the configuration variable q_i if and only if

$$\frac{\partial L}{\partial q_i} = 0, \quad i \in \{1,\ldots,n\}$$

Several underactuated systems possess certain symmetry property, for example the Lagrangian of a helicopters or of satellites is independent of their position, which gives rise to a symmetry (invariance of the Lagrangian).

Let us denote by p_i the ith generalized momentum defined by

$$p_i = \frac{\partial L}{\partial \dot{q}_i}$$

and consider the non-actuated Euler–Lagrange equation of motion:

$$\frac{d}{dt}\frac{\partial L}{\partial \dot{q}_i} - \frac{\partial L}{\partial q_i} = 0$$

An immediate consequence of the symmetry of the Lagrangian with respect to q_i is the conservation of the ith generalized momentum ($\dot{p}_i = 0$) and vice versa.

The equation $\dot{p}_i = 0$ is equivalent to constraints of first order,

$$W^T(q)\dot{q} = p_i(0)$$

where $W(q) = (m_{i1(q)}, \ldots, m_{in}(q))^T$ is the ith row of the inertia matrix $M(q)$. If this constraint is non-integrable, then the analysis of the system reduces to the analysis of a mechanical system with non-holonomic constraints of first order. Hence, the existence of symmetry leads to holonomic/non-holonomic velocity constraints for mechanical systems [23].

We shall employ a different notion of symmetry called kinetic symmetry with respect to q_i; that is,

$$\frac{\partial K}{\partial q_i} = 0$$

By definition, a mechanical system whose inertia matrix is independent of a set of configuration variables possesses a kinetic symmetry with respect to these variables. We call this subset: external variables and its complement: shape variables (the variables that appear in the inertia matrix). The kinetic symmetry is equivalent to the classical symmetry in the absence of potential energy.

The last part of this chapter is dedicated to the presentation of some UMSs obtained via the Lagrange formalism.

3.9 Examples of Underactuated Mechanical Systems

In this section, we are going to present some models of UMSs, the majority of which represent useful benchmarks for nonlinear control. These examples include the cart-inverted pendulum (or simply cart-pole) system, the sliding mass on cart, the Tora (Translational oscillator rotational actuator), the Acrobot, the Pendubot, and the ball and beam system. Each example will be treated briefly.

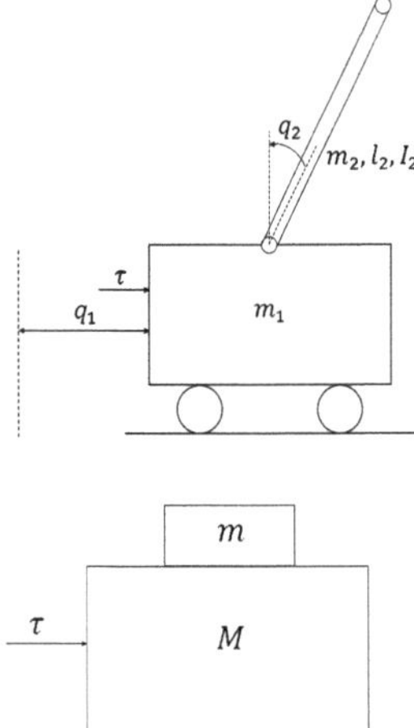

Fig. 3.2 The cart-inverted-pendulum system

Fig. 3.3 The sliding mass on cart system

3.9.1 The Cart-Inverted-Pendulum System

The cart-inverted pendulum is without doubt one of the most popular laboratory experiment used to illustrate nonlinear control techniques. The cart-inverted-pendulum system [2, 39], depicted in Fig. 3.2 is an UMS made of a cart that can move on a plane surface and a pendulum connected by a hinge on the cart. The overall system is controlled by an electric motor.

The necessity to control q_1 the cart displacement and q_2 the pendulum angular with a single control input τ classify this system under the class of UMSs.

The inertia matrix $M(q)$ and the potential energy $V(q)$ are given by

$$M(q) = \begin{bmatrix} m_1 + m_2 & m_2 l_2 \cos q_2 \\ m_2 l_2 \cos q_2 & m_2 l_2^2 + I_2 \end{bmatrix} \quad \text{and} \quad V(q) = m_2 g l_2 \cos q_2$$

where m_1 is the mass of the cart and m_2, l_2, I_2 are the mass, length from the center of mass and inertia of the pendulum, respectively.

The cart-inverted pendulum has been the subject of several linear and nonlinear control algorithms tests. Note that the segway is an extended version of the cart-inverted-pendulum technology.

3.9.2 The Sliding Mass on Cart System

The sliding mass on cart [29] is illustrated in Fig. 3.3. Assume that there is a friction coefficient B between the mass m and the cart of mass M. Denoting by x_1 the position of the mass m with respect to the cart and by x_2 the position of the cart, the equations of motion of this system are given by

$$m\ddot{x}_1 - B(\dot{x}_1 - \dot{x}_2) = 0$$

$$M\ddot{x}_2 + B(\dot{x}_1 - \dot{x}_2) = \tau$$

Fig. 3.4   Tora system

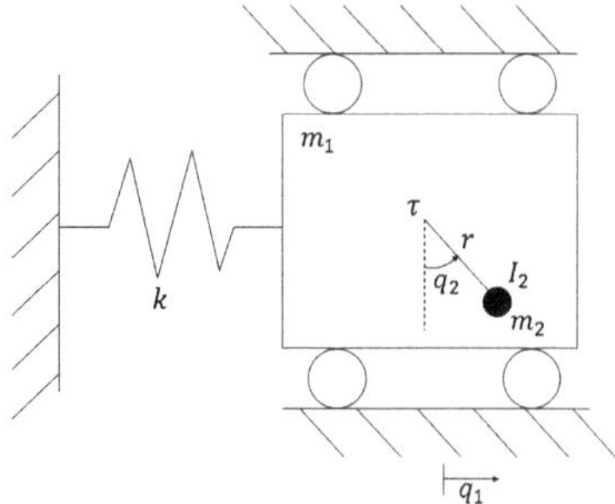

3.9.3 The Tora System

The Tora system[2] depicted in Fig. 3.4, is composed of an oscillatory platform which is controlled via a central mass [35]. The inertia matrix $M(q)$ and the potential energy $V(q)$ are given by

$$M(q) = \begin{bmatrix} m_1 + m_2 & m_2 r \cos q_2 \\ m_2 r \cos q_2 & m_2 r^2 + I_2 \end{bmatrix} \quad \text{and} \quad V(q) = \frac{1}{2} k_1 q_1^2 + m_2 g r \cos q_2$$

where q_1 is the platform displacement, q_2 is the pendulum angular, m_1 is the mass of the cart, m_2 the mass of the eccentric mass, r the radius of the rotation, k the spring constant, I_2 the inertia of the arm, g the gravity acceleration, and τ is the torque input.

3.9.4 The Acrobot and the Pendubot Systems

Consider a two-arm robot with a single control input. The actuation of the variable q_1 or q_2 yields two different UMSs: the Acrobot [7, 16], Fig. 3.5(a), where q_2 is actuated and the Pendubot [32], Fig. 3.5(b), whereby q_1 is actuated.

In reality, these two systems represent the same system when the latter is fully actuated. Any actuator failure or suppression of an actuator leads to two different structures. The inertia matrices of the two systems are given by

$$m_{11} = I_1 + I_2 + m_1 l_1^2 + m_2 \left(L_1^2 + l_2^2 \right) + 2 m_2 L_1 l_2 \cos q_2$$

$$m_{12} = m_{21} = I_2 + m_2 l_2^2 + m_2 L_1 l_2 \cos q_2$$

$$m_{22} = I_2 + m_2 l_2^2$$

and the corresponding potential energy $V(q)$ is given by

$$V(q) = (m_1 l_1 + m_2 L_1) g \cos q_1 + m_2 l_2 g \cos(q_1 + q_2)$$

---

[2]Translational Oscillator Rotational Actuator.

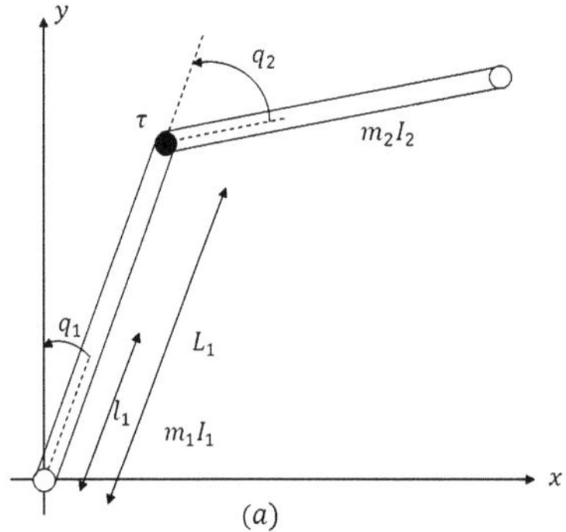

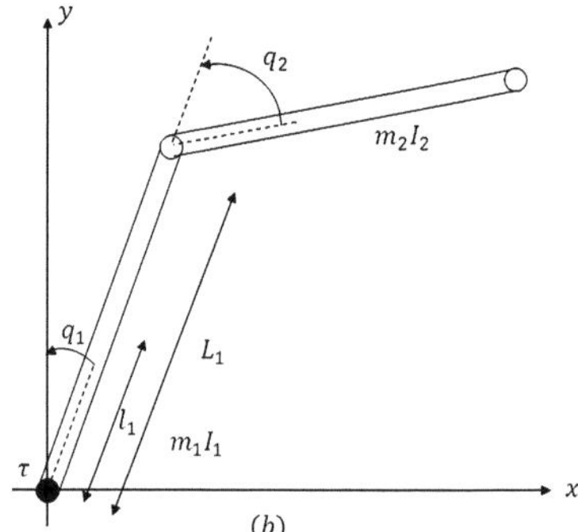

Fig. 3.5 Acrobot and Pendubot systems

Fig. 3.6 Inertia wheel
pendulum system

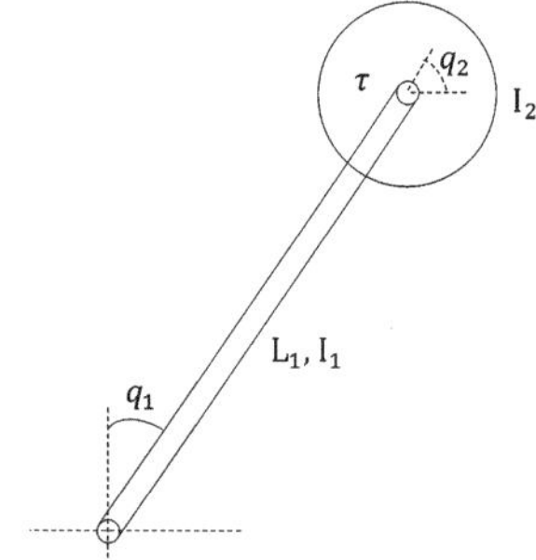

where m_i, I_i, L_i, l_i are the mass, inertia, length, and the length of the center of mass of the ith arm, respectively (see [10] for more details of the modeling of these systems).

3.9.5 The Inertial Wheel Pendulum System

The inertia or inertial wheel pendulum [6, 34] illustrated in Fig. 3.6 consists of a pendulum that has on its extremity a rotative inertial wheel. Here the pendulum is not actuated and the system is controlled via the wheel. The objective of the control is to firstly stabilize the pendulum at the vertical position and secondly stabilize the rotational movement of the wheel.

The entries of the inertia matrix associated with this system are given by

$$m_{11} = m_1 l_1^2 + m_2 L_1^2 + I_1 + I_2$$

$$m_{12} = m_{21} = m_{22} = I_2$$

The fact that the entries of this system are constant qualifies this system as flat.

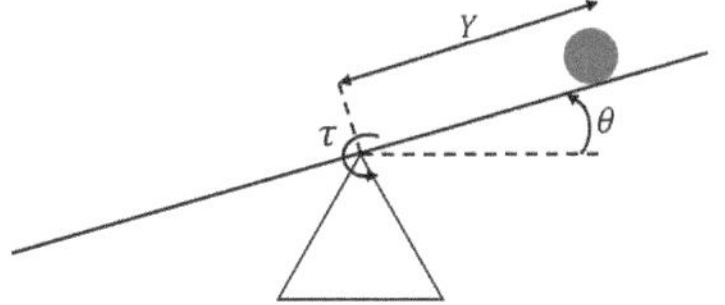

Fig. 3.7 Ball and beam system

The corresponding potential energy $V(q)$ is given by

$$V(q) = (m_1 l_1 + m_2 L_1) g \cos q_1 = m_0 \cos q_1$$

where m_1, m_2, I_1, I_2 are the masses, the inertia of the pendulum and the wheel, respectively; L_1, l_1 represent the length and the length of the center of the pendulum.

3.9.6 The Ball and Beam System

The ball and beam system [13] is composed of a beam that can pivot in the vertical plane via a torque τ at the center/point of rotation and a ball whose aim is to restrict its displacement to a sliding motion without friction along the beam Fig. 3.7.

In this example, the idea is to control the angular position θ of the beam and the position r of the ball by the only available control input τ applied at the center of the beam. The equations of motion are given by

$$\begin{cases} m\ddot{r} + mg\sin(\theta) - mr\dot{\theta}^2 = 0 \\ (mr^2 + I)\ddot{\theta} + 2mr\dot{r}\dot{\theta} + mgr\cos(\theta) = \tau \end{cases} \tag{3.16}$$

where I is the inertia of the beam, m the mass of the ball, and g the gravity acceleration.

3.10 Summary

In this chapter, we have presented the modeling of UMSs from the formalism of Lagrange. These systems are defined as having fewer control inputs than the degrees of freedom to be controlled. We have next explained the different problems associated to underactuation and then we have given the models of some examples of UMSs. We have highlighted the fact that these systems possess very few common properties so that they are studied on a case by case basis. In the next chapter we shall proceed to classify these systems with the aim to provide a generalized treatment of these systems or at least for some classes of these systems.

References

1. N.P. Aneke, Control of underactuated mechanical systems. Ph.D. thesis, Technishe Universiteit Eidhoven, 2003
2. K.J. Astrom, K. Furuta, Swining up a pendulum by energy control. Automatica **36**, 287–295 (2000)

3. M. Bennani, P. Rouchon, Robust stabilization of flat and chained systems, in *Proc. European Control Conf.* (1995), pp. 1781–1786
4. A.M. Bloch, J. Baillieul, *Nonholonomic Mechanics and Control* (Springer, Berlin, 2003)
5. A.M. Bloch, M. Reyhanoglu, N.H. McClamroch, Control and stabilization of nonholonomic dynamic systems. IEEE Trans. Autom. Control **37**(11), 1746–1757 (1992)
6. D.J. Block, K.J. Astrom, M.W. Spong, *The Reaction Wheel Pendulum* (Morgan and Claypool, San Rafael, 2007)
7. S. Bortoff, M.W. Spong, Pseudolinearization of the Acrobot using spline functions, in *Proc. IEEE Conf. on Decision and Control*, Tuscan (1992), pp. 593–598
8. R.W. Brockett, *Asymptotic Stability and Feedback Stabilization* (Birkhäuser, Basel, 1983)
9. J.M. Coron, B. d'Andréa-Novel, Smooth stabilizing time-varying control laws for a class of nonlinear systems, in *IFAC Nonlinear Control Systems Design Symp. (NOLCOS)* (1992), pp. 413–418
10. I. Fantoni, R. Lozano, *Nonlinear Control for Underactuated Mechanical Systems* (Springer, Berlin, 2002)
11. G.O. Gantmacher, *Lectures in Analytical Mechanics* (Mir, Moscow, 1970)
12. H. Goldstein, *Classical Mechanics* (Addison-Wesley, Reading, 1980)
13. J. Hauser, S. Sastry, P. Kokotović, Nonlinear control via approximate input-output linearization. IEEE Trans. Autom. Control **37**(3), 392–398 (1992)
14. E. Jarzebowska, Stabilizability and motion tracking conditions for mechanical nonholonomic control systems, in *Mathematical Problems in Engineering* (2007)
15. I. Kolmanovsky, N.H. Mcclamroch, Developments in nonholonomic control problems. IEEE Control Syst. Mag. **15**(6), 20–36 (1995)
16. A.D. Mahindrakar, R.N. Banavar, A swinging up of the Acrobot based on a simple pendulum strategy. Int. J. Control **78**(6), 424–429 (2005)
17. R.T. M'Closkey, R.M. Murray, Exponential stabilization of driftless nonlinear control systems using homogeneous feedback. IEEE Trans. Autom. Control **42**(5), 614–628 (1997)
18. J. Meisel, *Principles of Electromechanical Energy Conversion* (John Wiley and Sons, New York, 1959)
19. P. Morin, J.B. Pomet, C. Samson, Design of homogeneous time-varying stabilizing control laws for driftless controllable systems via oscillatory approximation of lie brackets in closed-loop. SIAM J. Control Optim. **38**, 22–49 (1997)
20. R.M. Murray, S. Sastry, Nonholonomic motion planning: steering using sinusoids. IEEE Trans. Autom. Control **38**(5), 700–716 (1993)
21. Y. Nakamura, W. Yoshihiko, O.J. Sordalen, Design and control of the nonholonomic manipulator. IEEE Trans. Robot. Autom. **17**(1), 48–59 (2001)
22. J.I. Neimark, F.A. Fufaev, *Dynamics of Nonholonomic Systems* (AMS, Providence, 1972)
23. R. Olfati Saber, Nonlinear control of underactuated mechanical systems with application to robotics and aerospace vehicles. Ph.D. thesis, Massachusetts Institute of Technology, Department Electrical Engineering and Computer Science, 2001
24. G. Oriolo, Y. Nakamura, Control of mechanical systems with second order nonholonomic constraints: underactuated manipulators, in *Proc. IEEE Conf. on Decision and Control*, UK (1991), pp. 2398–2403
25. J.B. Pomet, Explicit design of time-varying stabilizing control laws for a class of controllable systems without drift. Syst. Control Lett. **18**, 467–473 (1992)
26. M. Reyhanoglu, A. Van der Schaft, N.H. McClamroch, I. Kolmanovsky, Dynamics and control of a class of underactuated mechanical systems. IEEE Trans. Autom. Control **44**(9), 1663–1671 (1999)
27. M. Reyhanoglu, S. Cho, N.H. McClamroch, I. Kolmanovsky, Discontinuous feedback control of a planar rigid body with an unactuated internal degree of freedom, in *Proc. IEEE Conf. on Decision and Control*, Florida (1998), pp. 433–438
28. C. Samson, K. Ait-Abderrahim, Feedback control of a nonholonomic wheeled cart in Cartesian space, in *Proc. IEEE Conf. on Robotics and Automation* (1991), pp. 1136–1141

29. D. Seto, J. Baillieul, Control problems in super articulated mechanical systems. IEEE Trans. Autom. Control **39**(12), 2442–2453 (1994)
30. O. Sordalen, O. Egeland, Exponential stabilization of chained nonholonomic systems. IEEE Trans. Autom. Control **40**(1), 35–49 (2001)
31. M.W. Spong, *Underactuated Mechanical Systems* (Springer, Berlin, 1997)
32. M.W. Spong, D.J. Block, The Pendubot: a mechatronic system for control research and education, in *Proc. of the 34th IEEE Conf. on Decision and Control*, New Orleans (1995)
33. M.W. Spong, M. Vidyasagar, *Robot Dynamics and Control* (John Wiley and Sons, New York, 1989)
34. M.W. Spong, P. Corke, R. Lozano, Nonlinear control of the inertia wheel pendulum. Automatica **37**, 1845–1851 (1999)
35. C.J. Wan, D.S. Bernstein, V.T. Coppola, Global stabilization of the oscillating eccentric rotor. Nonlinear Dyn. **10**, 49–62 (1996)
36. J.T.Y. Wen, *Control of Nonholonomic Systems* (CRC Press/IEEE Press, Boca Raton/Los Alamitos, 1996)
37. D.C. White, H.H. Woodson, *Electromechanical Energy Conversion* (John Wiley and Sons, New York, 1959)
38. K.Y. Wichlund, O.J. Sordalen, O. Egeland, Control of vehicles with second-order nonholonomic constraints, in *Proc. European Control Conf.* (1995), pp. 3086–3091
39. H. Yu, Y. Liu, T. Yang, Tracking control of a pendulum-driven cart-pole underactuated system, in *Proc. IEEE Int. Conf. on Systems, Man and Cybernetics*, Japan (2007), pp. 2425–2430

Chapter 4
Classification of Underactuated Mechanical Systems

> *…order leads to all virtues but what leads to order?*
>
> *Georg Christoph Lichtenberg*

In order to provide a generalized method for the synthesis of control laws for under-actuated mechanical systems (UMSs), it is intuitive to try to look for the structural properties that are common to different UMSs. On the other hand, as has been mentioned in previous chapters, very often these systems are treated on a case by case basis. So far, only two attempts in classifying these systems are available in literature. The first classification is due to *Seto and Baillieul* and the second one is due to *Olfati-Saber*. The first objective of this chapter is to present these two classifications. The second objective is to investigate whether there is eventually a link between these two classifications.

4.1 Classification of UMSs According to *Seto and Baillieul*

One of the two available classifications of UMSs in the literature is due to Dambing Seto and John Baillieul [5]. For this, a graphical characteristic of UMSs is developed using the so-called control flow diagram (CFD), which is constructed to represent the interaction forces through the system's degrees of freedom. In doing so, three structures are identified: the chain structure, tree structure, and isolated vertex (or point) structure.

Seto and Baillieul, in their classification, give a control solution for UMSs having a chain structure whereby a systematic backstepping control design procedure is put in place. This is precisely the strong point of this classification. However, the weak point of this classification is that for the other two structures (tree and isolated vertex) the problem of deriving a systematic control design approach is still open.

For the rest of the discussion in this chapter, we shall refer to a mechanical system of n degrees of freedom described in the generalized coordinates $q = (q_1, q_2, \ldots, q_n)$ using the Lagrange formalism as explained in the previous chapter:

$$M(q)\ddot{q} + N(q, \dot{q}) = F(q)\tau \tag{4.1}$$

where $N(q, \dot{q}) = C(q, \dot{q})\dot{q} + G(q)$ according to (3.3).

A. Choukchou-Braham et al., *Analysis and Control of Underactuated Mechanical Systems*, DOI 10.1007/978-3-319-02636-7_4,
© Springer International Publishing Switzerland 2014

4.1.1 Principle of Control Flow Diagram

The main idea of the control flow diagram is to understand the relations that exist
between the elements of a system by studying the interacting forces that are coupled
with the degrees of freedom and to develop a graphical representation to capture
the dynamics of a given system. As mentioned before, the three structures chain,
tree, and isolated vertices are identified. Additionally, one can have a combination
of these three structures so that overall we have seven structures. These different
structures define the degree of complexity of a given system.

4.1.1.1 Graphical Characterization of UMSs

In this section, we recall the construction of the control flow diagram (CFD) associated to a given mechanical system and identify the different possible structures in
the CFD [5]. Such an analysis can be a starting point for the control synthesis of
UMSs.

For each point $(q^0, \dot{q}^0)$ and each neighborhood U of $(q^0, \dot{q}^0)$ we assign to system
(4.1) a digraph called CFD constructed as follows:

1. Rewrite system (4.1) in the form

$$\ddot{q} = M^{-1}(q)\big[F(q)\tau - N(q,\dot{q})\big] = H(q,\dot{q},\tau) \qquad (4.2)$$

2. Choose $n + m$ vertices $q_1, \ldots, q_n$ and $\tau_1, \ldots, \tau_m$.
3. For each vertex q_i, $i \in [1,n]$, draw a branch from τ_j and q_k $(j \in [1,m], k \in [1,n])$ with $k \neq i$, towards q_i if the function H_i depends explicitly on τ_j, q_k or $\dot{q}_k$ and associate the number a_{ij} to the branches between q_i and τ_j and the number b_{ik} to branches linking q_k to q_i.

   The values of a_{ij} and b_{ik} are determined as follows:

$$a_{ij} = \begin{cases} 2 & \text{if } \frac{\partial H_i}{\partial \tau_j} \neq 0 \text{ at } (q^0, \dot{q}^0) \\[2ex] -2 & \text{if } \frac{\partial H_i}{\partial \tau_j} \neq 0 \ \forall (q,\dot{q}) \in U \text{ except at } (q^0, \dot{q}^0) \end{cases}$$

$$b_{ik} = \begin{cases} 1 & \text{if } \frac{\partial H_i}{\partial \dot{q}_k} \neq 0 \text{ at } (q^0, \dot{q}^0) \\[2ex] -1 & \text{if } \frac{\partial H_i}{\partial \dot{q}_k} \neq 0 \ \forall (q,\dot{q}) \in U \text{ except at } (q^0, \dot{q}^0) \\[2ex] 2 & \text{if } \frac{\partial H_i}{\partial \dot{q}_k} \equiv 0 \ \forall (q,\dot{q}) \in U \text{ and } \frac{\partial H_i}{\partial q_k} \neq 0 \text{ at } (q^0, \dot{q}^0) \\[2ex] -2 & \text{if } \frac{\partial H_i}{\partial \dot{q}_k} \equiv 0 \ \forall (q,\dot{q}) \in U \text{ and } \frac{\partial H_i}{\partial q_k} \neq 0 \ \forall (q,\dot{q}) \in U \text{ except at } (q^0, \dot{q}^0) \end{cases}$$

4. Associate a length to each control path from τ_j to q_i $(i \in [1,n], j \in [1,m])$ by adding absolute values assigned to each branch. For each configuration vari- able q_i, keep the shortest control path, and among them eliminate all the singular paths (that is, the control paths containing branches with a negative number).

This completes the construction of the CFD.

Remark 4.1   It is also possible to construct the CFD of UMSs under first order non-holonomic constraints, for more details refer to [5].

When a mechanical system is subjected to first order non-holonomic constraints (Sect. 3.4), its dynamics can be written as follows:

$$M(q)\ddot{q} + N(q,\dot{q}) = W(q)\lambda + F(q)\tau$$
$$W^T(q)\dot{q} = 0 \tag{4.3}$$

where the term $W(q)\lambda$ can be considered as a necessary force to maintain the constraints. Such forces can be eliminated by studying a family of differential equations of lower dimension, which represents the system evolution on the constraints manifold and in which the system appears holonomic.

Without loss of generality, the configuration variables are divided into two sets $q_a = \{q_1, \ldots, q_d\}$ and $q_b = \{q_{d+1}, \ldots, q_n\}$, and the matrix $W(q)$ into two submatrices $W_a(q)$ and $W_b(q)$, of dimensions $(n \times d) \times d$ and $(n-d) \times (n-d)$, respectively, such that $\det(W_b(q)) \neq 0$.

If we define $T(q) = \begin{bmatrix} I \\ -(W_b^{-1}W_a) \end{bmatrix}$ then $W(q)T(q) = 0$.

In this case, one can write (see [1] for more details):

$$\ddot{q}_a = \left[T^T(q)M(q)T(q)\right]^{-1}T^T(q)\left[F(q)\tau - N\left(q, T(q)\dot{q}_a\right) - M(q)\dot{T}(q)\dot{q}_a\right]$$
$$\dot{q}_b = -\left(W_b^{-1}(q)W_a(q)\right)^T \dot{q}_a \tag{4.4}$$

where the term $W(q)\lambda$ is eliminated.

Remark 4.2   The d independent vector fields represented by the columns of the matrix T generate the null space of the matrix $W(q)$. By multiplying Eq. (4.3) by T^T (which is done to obtain (4.4)), the dynamics of the system are projected towards the null space of $W(q)$. Consequently, the term $W^T(q)$ is eliminated. This reduction is not unique, different partitions of the variables q_a and q_b lead to different representations for (4.4). Moreover, the reduction presented is only local and a theory for global reduction for non-holonomic systems remains to be developed.

Proof   We have

$$W^T(q)\dot{q} = 0 \quad \Rightarrow \quad \begin{bmatrix} W_a^T & W_b^T \end{bmatrix}\begin{bmatrix} \dot{q}_a \\ \dot{q}_b \end{bmatrix} = 0$$

which gives

$$\dot{q}_b = -\left(W_a(q)W_b^{-1}(q)\right)^T \dot{q}_a \tag{4.5}$$

Hence

$$\dot{q} = \begin{bmatrix} \dot{q}_a \\ \dot{q}_b \end{bmatrix} = \begin{bmatrix} \dot{q}_a \\ -(W_a(q)W_b^{-1}(q))^T \dot{q}_a \end{bmatrix} = T(q)\dot{q}_a \quad \Rightarrow \quad \ddot{q} = \dot{T}(q)\dot{q}_a + T(q)\ddot{q}_a$$

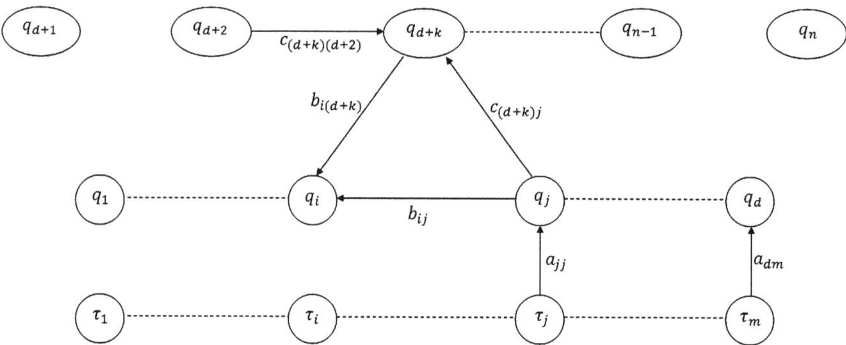

Fig. 4.1 Construction of a CFD in the general case

Let us multiply (4.3) by $T^T(q)$ and replace $\ddot{q}$ by $\ddot{q} = \dot{T}(q)\dot{q}_a + T(q)\ddot{q}_a$, then

$$T^T M(q)\big(\dot{T}(q)\dot{q}_a + T(q)\ddot{q}_a\big) + T^T N(q,\dot{q}) = T^T W(q)\lambda + T^T F(q)\tau \qquad (4.6)$$

This yields (4.4). $\square$

The CFD for first order non-holonomic systems is obtained as follows:

1. Rewrite (4.3) under the form (4.4) and set $\ddot{q}_a = H_a(q,\dot{q}_a,\tau)$ and $\dot{q}_b = H_b(q,\dot{q}_a)$ where $H_a = [H_a^1,\ldots,H_a^d]^T$ and $H_b = [H_b^{d+1},\ldots,H_b^n]^T$.
2. Choose $n+m$ vertices $q_1,\ldots,q_n$ and $\tau_1,\ldots,\tau_m$.
3. For each q_i, $i = 1,\ldots,d$, repeat step 3 for the holonomic systems with H_i replaced by H_a^i.
4. For each q_i, $i \in [d+1,\ldots,n]$, draw a branch from q_k, $k = 1,\ldots,n$ $(k \neq i)$, towards q_i if the function H_b^i depends explicitly on q_k or $\dot{q}_k$, and associate the number c_{ik} to each branch that is determined by

$$c_{ik} = \begin{cases} 0 & \text{if } \frac{\partial H_b^i}{\partial \dot{q}_k} \neq 0 \text{ at } (q^0,\dot{q}^0) \\[2ex] -0 & \text{if } \frac{\partial H_b^i}{\partial \dot{q}_k} \neq 0 \; \forall (q,\dot{q}) \in U \text{ except at } (q^0,\dot{q}^0) \\[2ex] 1 & \text{if } \frac{\partial H_b^i}{\partial \dot{q}_k} \equiv 0 \; \forall (q,\dot{q}) \in U \text{ and } \frac{\partial H_b^i}{\partial q_k} \neq 0 \text{ at } (q^0,\dot{q}^0) \\[2ex] -1 & \text{if } \frac{\partial H_b^i}{\partial \dot{q}_k} \equiv 0 \; \forall (q,\dot{q}) \in U \text{ and } \frac{\partial H_b^i}{\partial q_k} \neq 0 \; \forall (q,\dot{q}) \in U \text{ except at } (q^0,\dot{q}^0) \end{cases}$$

5. Same as step 4 of the previous organigram.

Remark 4.3 The formal notation -0 is adopted to distinguish the case in which the differential-algebraic constraint in (4.3) has singularities at $(q^0,\dot{q}^0)$.

A part of the CFD for first order non-holonomic systems is shown in Fig. 4.1 [5].

Let us now define the three different structures: chain structure, tree structure, and isolated vertices structure.

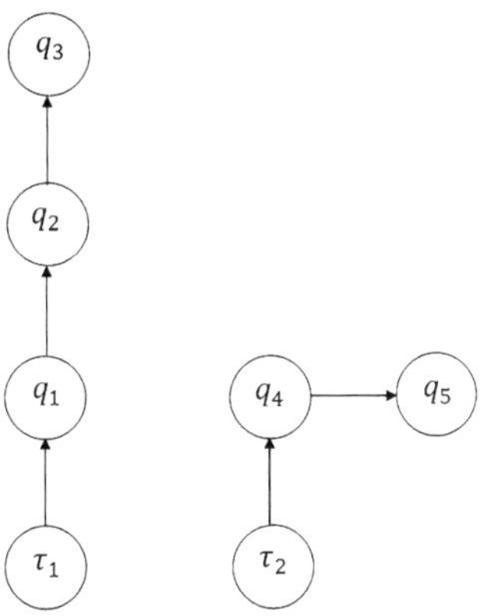

Fig. 4.2 Chain structures

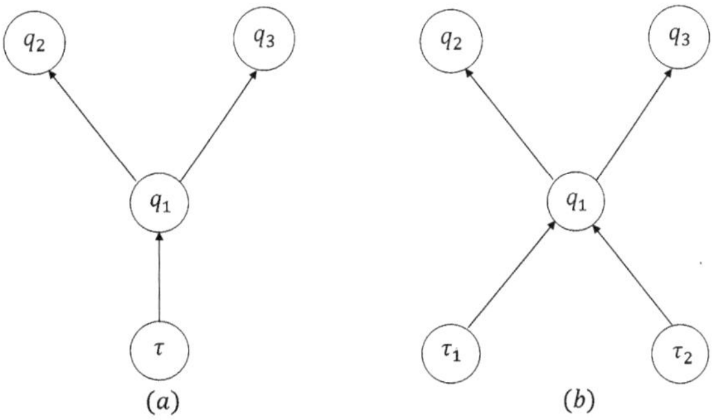

Fig. 4.3 Tree structures

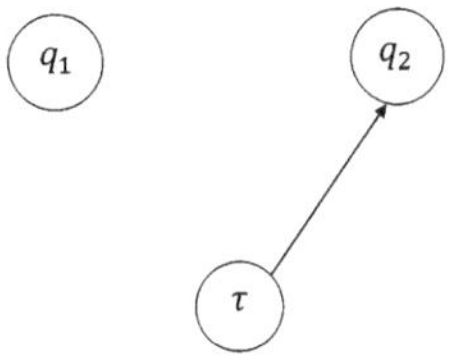

Fig. 4.4 Isolated vertex
structure

Definition 4.1 [5] Suppose that G_i is a subgraph of the CFD, and that G_i contains m_i control inputs.

1. G_i has a chain structure if there are m_i vertices with m_i control path covering all the vertices in G_i, such that each vertex belongs to one and only one control path, see Fig. 4.2.
2. G_i has a tree structure if for any m_i vertices, every corresponding set of m_i control paths will either leave some vertices in m_i not covered Fig. 4.3(a) or cover all the vertices in m_i with some vertices appearing on more than one path Fig. 4.3(b).
3. G_i has an isolated vertices structure if G_i only contains vertices to which there are only singular control paths or no control paths in the CFD Fig. 4.4.

Based on these definitions, *Seto and Baillieul* defined the notion of degree of complexity in the context of CFD. This degree indicates the difficulty in controlling a system.

Definition 4.2 [5] By the degree of complexity of a given system, we mean the position of the system in the following hierarchy:

$$
\begin{cases}
(1) \ \text{chains only} \\
(2) \ \text{chains} \ + \ \text{trees} \\
(3) \ \text{trees only} \\
(4) \ \text{chains} \ + \ \text{isolated vertices} \\
(5) \ \text{chains} \ + \ \text{trees} \ + \ \text{isolated vertices} \\
(6) \ \text{trees} \ + \ \text{isolated vertices} \\
(7) \ \text{isolated vertices only}
\end{cases}
$$

Remark 4.4

- Clearly, the chain structure is the least difficult to control, it even appears that systems with such structure can be controlled via feedback linearization or by backstepping since the degrees of freedom and the control are connected in series.
- On the other hand, systems with tree structure, do not have the same advantages, because we have to control certain degrees of freedom in parallel; in the sense that one control input has to control more than one variable simultaneously, which limits the control objectives.
- For systems with isolated vertices structure, certain control objectives are difficult or even impossible to attain because the control does not have any influence on some variables.

4.1.1.2 Interpretation of CFD

Let us define an output function $y_i = q_i$, $i = 1, \dots, n$, and the relative degree for each output y_i.

Definition 4.3 [5] The relative degree for a configuration variable (if it exists) is the shortest length of the control path leading to this variable.

It is clear that the configuration variables in an isolated vertices structure do not have a well-defined relative degree.

Considering the relative degree r_k of q_k, $k \in [1, n]$, the controllability of q_k is summarized in Table 4.1

4.1.2 Examples

To illustrate this concept, we have constructed the CFDs of some UMSs.

Table 4.1 Controllability of configuration variables in a CFD

r_k is well defined at $(q^0, \dot{q}^0)$	q_k is directly controllable by $\tau \in U : r_k = 2$ q_k is indirectly controllable by $\tau \in U : r_k > 2$
r_k is not well defined at $(q^0, \dot{q}^0)$	q_k is in a singular control path and is not affected by $\tau \in U$ at $(q^0, \dot{q}^0)$ q_k is not in control path and cannot be controlled

Fig. 4.5 CFD of the sliding mass on cart system

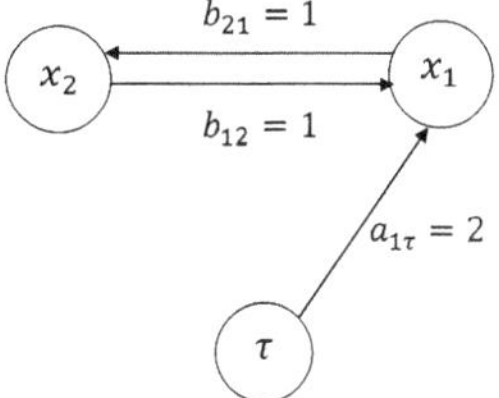

1. Consider the mechanical system consisting of a sliding mass on a cart as depicted in Fig. 3.3. The equations of motion are given by

$$\ddot{x}_1 = \frac{B}{m}(\dot{x}_1 - \dot{x}_2) = H_1(\dot{x}_1, \dot{x}_2)$$

$$\ddot{x}_2 = \frac{1}{M}\left[\tau - B(\dot{x}_1 - \dot{x}_2)\right] = H_2(\dot{x}_1, \dot{x}_2, \tau)$$

The function H_1 does not depend on the input τ, hence the path linking τ to x_1 does not exist. However, there is an indirect path linking τ to x_1 through x_2 of length (relative degree) $a_{2\tau} + b_{12} = 3$ Fig. 4.5; the values of $a_{2\tau}$, b_{21}, and b_{12} are determined by

$$\frac{\partial H_2}{\partial \tau} = \frac{1}{M} \neq 0 \quad \text{in } \left(x^0, \dot{x}^0\right) \quad \Rightarrow \quad a_{2\tau} = 2$$

$$\frac{\partial H_1}{\partial \dot{x}_2} = -\frac{B}{m} \neq 0 \quad \text{in } \left(x^0, \dot{x}^0\right) \quad \Rightarrow \quad b_{12} = 1$$

$$\frac{\partial H_2}{\partial \dot{x}_1} = -\frac{B}{M} \neq 0 \quad \text{in } \left(x^0, \dot{x}^0\right) \quad \Rightarrow \quad b_{21} = 1$$

The sliding mass on cart system possesses a chain structure. The variable x_2 is directly controllable by τ, while the variable x_1 is controllable via x_2.

Remark 4.5 The only examples of UMSs having a chain structure that can be found in literature are the sliding mass on cart and the robotic arm with joint elasticity [2].

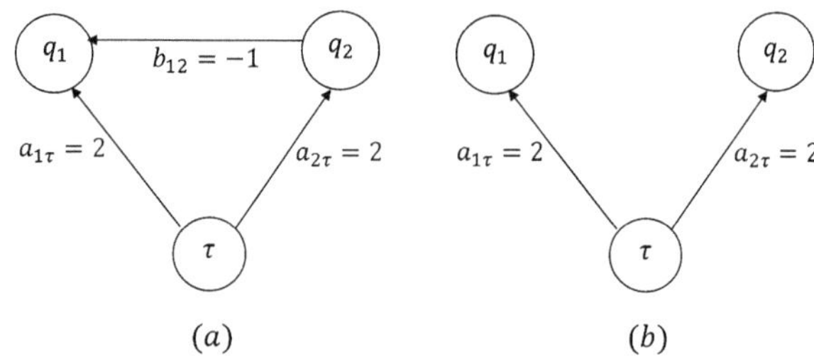

Fig. 4.6 CFD of the inverted pendulum system

2. Consider the inverted pendulum on a cart as depicted on Fig. 3.2; the equations of motion are given by

$$\ddot{q}_1 = \frac{1}{\det(M)}\left[(m_2 l_2^2 + I_2)\tau + (m_2 l_2^2 + I_2)m_2 l_2 \dot{q}_2^2 \sin q_2 - m_2^2 l_2^2 g \sin q_2\right]$$

$$= H_1(q_2, \dot{q}_2, \tau)$$

$$\ddot{q}_2 = \frac{1}{\det(M)}\left[(-m_2 l_2 \cos q_2)\tau + (m_1 + m_2)m_2 g l_2 \sin q_2 - m_2^2 l_2^2 \dot{q}_2^2 \sin q_2\right]$$

$$= H_2(q_2, \dot{q}_2, \tau)$$

where $\det(M)$ is the determinant of the inertia matrix $M(q)$ as given previously. The values of $a_{1\tau}$, $a_{2\tau}$, and b_{12} are determined by

$$\frac{\partial H_1}{\partial \tau} = \frac{m_2 l_2^2 + I_2}{\det(M)} \neq 0 \quad \text{in } (q^0, \dot{q}^0) \quad \Rightarrow \quad a_{1\tau} = 2$$

$$\frac{\partial H_2}{\partial \tau} = -\frac{m_2 l_2 \cos q_2}{\det(M)} \neq 0 \quad \text{in } (q^0, \dot{q}^0) \quad \Rightarrow \quad a_{2\tau} = 2$$

$$\frac{\partial H_1}{\partial \dot{q}_2} = \frac{2(m_2 l_2^2 + I_2)m_2 l_2 \sin q_2}{\det(M)}\dot{q}_2 \neq 0 \quad \forall (q, \dot{q}) \in U \text{ except at } (q^0, \dot{q}^0)$$

$$\Rightarrow \quad b_{12} = -1$$

The function H_2 does not depend on q_1 or $\dot{q}_1$. Hence, the path linking q_1 to q_2 (b_{21}) does not exist Fig. 4.6(a).

There are two control paths linking the input variable τ to the variable q_1: the first is a direct path of length ($a_{1\tau} = 2$) and the second one is an indirect path passing through the variable q_2 of length ($a_{2\tau} + | b_{12} | = 3$). In this case, we keep the first control path of shortest length. Moreover, the second path is singular and therefore must be eliminated. The final CFD is given in Fig. 4.6(b). This system possesses a tree structure. Hence, it is necessary to control the two variables q_1 and q_2 simultaneously.

Remark 4.6 The Acrobot, the Pendubot, and the Tora system are also UMSs having a CFD in tree structure.

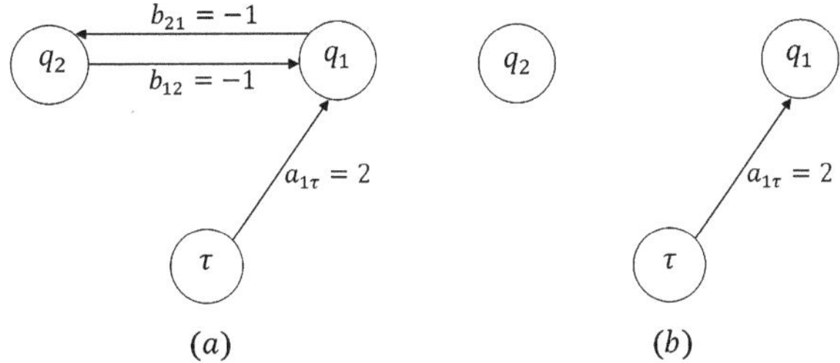

Fig. 4.7 CFD of the ball and beam system

3. Consider the ball and beam system depicted in Fig. 3.7. The equations of motion are given by

$$\ddot{q}_1 = \frac{1}{I + mq_2^2}(\tau - 2mq_2\dot{q}_1\dot{q}_2 - mgq_2\cos q_1) = H_1(q,\dot{q},\tau)$$

$$\ddot{q}_2 = \frac{1}{m}\left(mq_2\dot{q}_1^2 - mg\sin q_1\right) = H_2(q,\dot{q})$$

where θ is replaced by q_1 and r by q_2. The values of $a_{1\tau}$, $a_{2\tau}$, and b_{12} are

$$\frac{\partial H_1}{\partial \tau} = \frac{1}{I + mq_2^2} \neq 0 \quad \text{in } \left(q^0,\dot{q}^0\right) \quad \Rightarrow \quad a_{1\tau} = 2$$

$$\frac{\partial H_2}{\partial \dot{q}_1} = 2q_2\dot{q}_1 \neq 0 \quad \forall (q,\dot{q}) \in U \text{ except at } \left(q^0,\dot{q}^0\right) \quad \Rightarrow \quad b_{21} = -1$$

$$\frac{\partial H_1}{\partial \dot{q}_2} = \frac{-2m}{I + mq_2^2}q_2\dot{q}_1 \neq 0 \quad \forall (q,\dot{q}) \in U \text{ except at } \left(q^0,\dot{q}^0\right) \quad \Rightarrow \quad b_{12} = -1$$

Figure 4.7(a) represents the CFD of the ball and beam example. Note that q_1 is linked to q_2 via a centrifugal force $mq_2\dot{q}_1^2$. In this case, the relative degree of q_2 with respect to τ is not defined when $\dot{q}_1 = 0$. Figure 4.7(b) shows the final CFD (after eliminating the singular control path) which is an isolated vertex structure.

According to this classification, we conclude that the degree of complexity to control the systems of the above examples increases in the following order: the sliding mass on cart system, next the inverted pendulum system, and last the ball and beam system.

4.2 Classification of UMSs According to *Olfati-Saber*

The second classification of UMSs is due to *Reza Olfati-Saber*. His main contribution was to determine an explicit change of coordinates that transforms the coupled subsystems obtained from the partial linearization of Spong into uncoupled cascaded systems in normal form. The resulting normal forms leads to a second classification of UMSs based on structural properties.

The advantage of such a classification is that it allows to define a suitable control, according to the obtained class. For instance, the control of a system that can be put into a normal strict feedback form would be done by a backstepping procedure; while those systems having a feedforward normal form would be done via a forwarding scheme. On the other hand, for those systems having a non-triangular form, the control problem for the latter is still an open issue (except in some particular cases).

4.2.1 Normal Forms of UMSs

According to the previous chapter, one can always partially linearize the dynamics of UMSs. However, the new control appears in the two subsystems: linear (q_2, p_2) and nonlinear (q_1, p_1), that is,

$$\dot{q}_1 = p_1$$
$$\dot{p}_1 = f_0(q, p) + g_0(q)u$$
$$\dot{q}_2 = p_2$$
$$\dot{p}_2 = u$$

whereby the new control input u is given by $\tau = \alpha(q)u + \beta(q, \dot{q})$.

The idea is to decouple these two subsystems via a global change of coordinates.

Theorem 4.1 [3] *Consider an UMS with inertia matrix $M(q) = \{m_{ij}(q)\}$; $i, j = 1, 2$ where $q = (q_1, q_2)$, $q_1 = (q_1^i) \in \mathbb{R}^{n-m}$, and $q_2 = (q_2^j) \in \mathbb{R}^m$ denote the actuated and non-actuated configuration variables, respectively. Let us define*

$$g(q) = \begin{bmatrix} g_0(q) \\ I_{m \times m} \end{bmatrix}$$

where $g_0(q) = -m_{11}^{-1}(q)m_{12}(q) = (g_0^1(q), \ldots, g_0^m(q))$ with $g_0^j(q) \in \mathbb{R}^n$, $j = 1, \ldots, m$ and $I_{m \times m}$ is the identity matrix.

Let us define the column-wise full rank distribution $\Delta(q)$ (globally non-singular).

$$\Delta(q) = \text{span}\{column\ of\ g(q)\}$$

Then, a necessary and sufficient condition for the distribution $\Delta(q)$ to be globally involutive; that is, completely integrable, is that

$$\frac{\partial g_0^j(q)}{\partial q_1} g_0^i(q) - \frac{\partial g_0^i(q)}{\partial q_1} g_0^j(q) + \frac{\partial g_0^j(q)}{\partial q_2^i} - \frac{\partial g_0^i(q)}{\partial q_2^j} = 0, \quad \forall i, j = 1, \ldots, m \quad (4.7)$$

In addition, if the condition (4.7) is verified, then there exists a global change in coordinates given by

$$z_1 = \Phi(q_1, q_2)$$
$$z_2 = \left(D_{q_1}\Phi(q)\right) \cdot p_1 + \left(D_{q_2}\Phi(q)\right) \cdot p_2$$
$$\xi_1 = q_2$$
$$\xi_2 = \dot{q}_2$$

that transforms the dynamics of the system in a normal form

$$\begin{aligned}
\dot{z}_1 &= z_2 \\
\dot{z}_2 &= f(z, \xi_1, \xi_2) \\
\dot{\xi}_1 &= \xi_2 \\
\dot{\xi}_2 &= u
\end{aligned} \tag{4.8}$$

Remark 4.7 The normal form (4.8) is a special case of the well-known Byrnes–Isidori form with double integrators

$$\begin{aligned}
\dot{z} &= f(z, \xi_1, \xi_2) \\
\dot{\xi}_1 &= \xi_2 \\
\dot{\xi}_2 &= u
\end{aligned} \tag{4.9}$$

Once decoupled, the system can take one of the cascaded normal forms as given below [3].

Definition 4.4 (Cascaded system)   A nonlinear system is said to be in the cascaded form if it has the following structure:

$$\begin{aligned}
\dot{z} &= f(z, \xi) \\
\dot{\xi} &= g(\xi, u)
\end{aligned} \tag{4.10}$$

where $f : \mathbb{R}^n \times \mathbb{R}^m \rightarrow \mathbb{R}^n$, $g : \mathbb{R}^m \times \mathbb{R}^p \rightarrow \mathbb{R}^m$, (z, ξ) is the composite state and u is the control input. If $\dot{\xi} = A\xi + Bu$, then the system (4.10) is said to be a partially linear cascade nonlinear system.

Definition 4.5 (Feedback form)   A nonlinear system is said to be in the strict feedback form if it possesses the following triangular structure:

$$\begin{aligned}
\dot{z} &= f(z, \xi_1) \\
\dot{\xi}_1 &= \xi_2 \\
&\;\;\vdots \\
\dot{\xi}_m &= u
\end{aligned}$$

Definition 4.6 (Feedforward form) A nonlinear system is said to be in the feedforward form if it possesses the following triangular structure:

$$\dot{x}_1 = x_2 + \varphi_1(x_2, \ldots, x_n, u)$$
$$\dot{x}_2 = x_3 + \varphi_2(x_3, \ldots, x_n, u)$$
$$\vdots$$
$$\dot{x}_n = u + \varphi_n(x_n, u)$$

Definition 4.7 (Non-triangular form) A nonlinear system is said to have a non-triangular form if it possesses the following structure:

$$\dot{z} = f(z, \xi_1, \xi_2, \ldots, \xi_m)$$
$$\dot{\xi}_1 = \xi_2$$
$$\vdots$$
$$\dot{\xi}_m = u$$

$$(4.11)$$

where $z \in \mathbb{R}^n$ and where $u \in \mathbb{R}^p$. The system (4.11) is also called normal form of Byrnes–Isidori.

Definition 4.8 (Non-triangular linear-quadratic form) A nonlinear system is said to have a non-triangular linear-quadratic form if it possesses the following structure:

$$\dot{z}_1 = \mu(z_1)z_2 + \eta(\xi_1)\xi_2$$
$$\dot{z}_2 = \phi(z_1, \xi_1) + \Sigma(\xi_1, z_2, \xi_2)$$
$$\dot{\xi}_1 = \xi_2$$
$$\dot{\xi}_2 = u$$

$$(4.12)$$

where $z_1, z_2 \in \mathbb{R}^n$, $u \in \mathbb{R}^p$, $\mu(z_1)$ is a positive definite matrix, $\phi : \mathbb{R}^n \times \mathbb{R}^p \to \mathbb{R}^n$ and $\Sigma : \mathbb{R}^p \times \mathbb{R}^n \times \mathbb{R}^p \to \mathbb{R}^n$ possesses a quadratic structure in (z_2, ξ_2)

$$\Sigma(\xi_1, z_2, \xi_2) = \begin{bmatrix} z_2 \\ \xi_2 \end{bmatrix}^T \Pi(\xi_1) \begin{bmatrix} z_2 \\ \xi_2 \end{bmatrix}$$

where $\Pi = (\Pi^1, \ldots, \Pi^n)$ is a cubic matrix with entries in $\mathbb{R}^{n \times n}$ and for $v \in \mathbb{R}^n$, $v^T \Pi v := (v^T \Pi^1 v, \ldots, v^T \Pi^n v)^T \in \mathbb{R}^n$. If $\eta \equiv 0$, (4.12) is called non-triangular quadratic form (with respect to ξ_2) . If $\Sigma \equiv 0$, (4.12) is called non-triangular linear form (with respect to ξ_2).

Olfati-Saber divided the UMSs into eight classes based on the obtained normal forms. We shall start by presenting the classification of UMSs with two degrees of freedom and then briefly recall the classification of UMSs of higher order.

4.2.2 UMSs with Two Degrees of Freedom

These systems can only give three different classes denoted Class I, Class II, and Class III. The considered systems are those which possess a kinetic symmetry with respect to q_1; that is $\frac{\partial K}{\partial q_1} = 0$ where K is the kinetic energy. In other words $M(q) = M(q_2)$.

Because of this property, the inertia matrix depends on a certain configuration variable q_2, called shape variable and is not dependent on the variable q_1 called external variable.

The general model for UMSs with two degrees of freedom is of the form [3]

$$m_{11}(q_2)\ddot{q}_1 + m_{12}(q_2)\ddot{q}_2 + m'_{11}(q_2)\dot{q}_1\dot{q}_2 + m'_{12}(q_2)\dot{q}_2^2 - g_1(q_1, q_2) = \tau_1$$

$$m_{21}(q_2)\ddot{q}_1 + m_{22}(q_2)\ddot{q}_2 - \frac{1}{2}m'_{11}(q_2)\dot{q}_1^2 + \frac{1}{2}m'_{22}(q_2)\dot{q}_2^2 - g_2(q_1, q_2) = \tau_2$$

where $g_i(q_1, q_2) = -\partial V(q)/\partial q_i$, $i = 1, 2$, and $'$ denotes d/dq_2.

Class I are those for which q_2 is actuated ($\tau_1 = 0$). Class II are those for which q_2 is not actuated ($\tau_2 = 0$).

It is shown that every underactuated system of Class I can be transformed into a strict feedback form.

Proposition 4.1 ([3] (Class I)) *The global change of coordinates (obtained from the Lagrangian) given by the following equations*:

$$\begin{cases} z_1 = q_1 + \gamma(q_2) \\ z_2 = m_{11}(q_2)p_1 + m_{12}(q_2)p_2 = \frac{\partial L}{\partial \dot{q}_1} \\ \xi_1 = q_2 \\ \xi_2 = p_2 \end{cases} \tag{4.13}$$

where $\gamma(q_2) = \int_0^{q_2} m_{11}^{-1}(\theta)m_{12}(\theta)\, d\theta$ transforms the dynamics of the system into a cascaded strict feedback form

$$\begin{aligned} \dot{z}_1 &= m_{11}^{-1}(\xi_1)z_2 \\ \dot{z}_2 &= g_1\big(z_1 - \gamma(\xi_1), \xi_1\big) \\ \dot{\xi}_1 &= \xi_2 \\ \dot{\xi}_2 &= u \end{aligned} \tag{4.14}$$

such that u is the new control obtained from collocated partial feedback linearization and $g_1(q_1, q_2) = -\partial V(q)/\partial q_1$.

Corollary 4.1 *The Acrobot and the Tora are Class I UMSs with two degrees of freedom that can be transformed into a strict feedback form.*

Proposition 4.2 ([3] (Class II)) *The explicit change of coordinates (obtained from the Lagrangian) given by the following equations:*

$$\begin{cases} z_1 = q_1 + \gamma(q_2) \\ z_2 = m_{21}(q_2)p_1 + m_{22}(q_2)p_2 = \frac{\partial L}{\partial \dot{q}_2} \\ \xi_1 = q_2 \\ \xi_2 = p_2 \end{cases} \tag{4.15}$$

where $\gamma(q_2) = \int_0^{q_2} m_{21}^{-1}(\theta)m_{22}(\theta)\,d\theta$ is defined in the set $U = \{q_2 / m_{21}(q_2) \neq 0\}$ transforms the dynamics of the system into a cascaded non-triangular quadratic form:

$$\begin{aligned} \dot{z}_1 &= m_{21}^{-1}(\xi_1)z_2 \\ \dot{z}_2 &= g_2\big(z_1 - \gamma(\xi_1), \xi_1\big) \\ &\quad + \frac{m'_{11}(\xi_1)}{2m_{21}^2(\xi_1)}z_2^2 + \left\{ \frac{m'_{21}(\xi_1)}{m_{21}(\xi_1)} - \frac{m_{22}(\xi_1)m'_{11}(\xi_1)}{2m_{21}^2(\xi_1)} \right\}z_2\xi_2 \\ &\quad + \left\{ \frac{m_{22}^2(\xi_1)}{2m_{21}^2(\xi_1)}m'_{11}(\xi_1) - \frac{m_{22}(\xi_1)}{m_{21}(\xi_1)}m'_{21}(\xi_1) + \frac{1}{2}m'_{22}(\xi_1) \right\}\xi_2^2 \\ \dot{\xi}_1 &= \xi_2 \\ \dot{\xi}_2 &= u \end{aligned} \tag{4.16}$$

such that u is the new control input obtained by non-collocated partial linearization.

Corollary 4.2 *The inverted pendulum, the ball and beam and the Pendubot are Class II underactuated systems with two degrees of freedom. They can be transformed into a non-triangular quadratic form on the set $U = \{q_2 / m_{21}(q_2) \neq 0\}$.*

Proposition 4.3 ([3] (Class III)) *Consider Class-II systems and assume that the following supplementary conditions are satisfied:*

(i) $g_2(q_1, q_2)$ *is not dependent on q_1; that is, $D_{q_1} D_{q_2} V(q) = 0$*
(ii) m_{11} *is constant*
(iii) $\psi(q_2) = g_2(q_2)/m_{21}(q_2)$ *satisfies $\psi'(0) \neq 0$ then the change of coordinates $y_1 = z_1$, $y_2 = m_{21}^{-1}(\xi_1)z_2$ transforms the non-triangular form* (4.16) *into the feedforward form*

$$\begin{aligned} \dot{y}_1 &= y_2 \\ \dot{y}_2 &= \psi(\xi_1) + \left\{ \frac{m'_{22}(\xi_1)}{2m_{21}(\xi_1)} - \frac{m_{22}(\xi_1)}{m_{21}^2(\xi_1)}m'_{21}(\xi_1) \right\}\xi_2^2 \\ \dot{\xi}_1 &= \xi_2 \\ \dot{\xi}_2 &= u \end{aligned} \tag{4.17}$$

Corollary 4.3 *The inverted cart-pendulum system satisfies the three conditions of Proposition* (4.3) ($\psi(q_2) = g \tan(q_2) \Rightarrow \psi'(0) = g \neq 0$). *Consequently, the system is transformable into a feedforward form* (4.17).

The above considered systems are UMSs with two degrees of freedom. However, real systems, which have a definite physical meaning, are those UMSs with higher degrees of freedom. It is therefore necessary to give the Olfati-Saber's classification of these systems as well. Nonetheless, in order to not overload this presentation, we shall try to be very brief. For more details on this subject see [3] and [4].

4.2.3 Classification of High-Order UMSs

The main idea here is to reduce high-order nonlinear UMSs into a nonlinear low-order subsystem cascaded with a linear subsystem. The obtained normal forms give rise to the second classification. At the same time, the control design for the nonlinear subsystem is less complicated but must be extended to the overall system through backstepping or forwarding procedures according to the obtained normal forms.

As for two degrees of freedom UMSs, we consider UMSs with kinetic symmetry where the shape variables that appear in the inertia matrix are denoted by q_s, while the external variables that do not appear in the inertia matrix are denoted by q_x.

The exploitation of such property has permitted to reduce the complexity of the control synthesis for UMSs. In effect, under some change of coordinates, the initial system is transformed into two cascaded subsystems where the first subsystem is nonlinear and the second one is linear, often in the form of a chain of integrators.

The analytical tools allowing such reduction are the generalized momenta and their integrals computed from the Lagrangian. Nevertheless, several benchmarks and real systems does not possess integrable generalized momentums. Then, these momenta are decomposed into a sum of an integrable part and a non-integrable part, which is considered as a perturbation of the integrable case and will appear only in the reduced nonlinear subsystem.

The Euler–Lagrange equations for these systems are given by

$$m_{xx}(q_s)\ddot{q}_x + m_{xs}(q_s)\ddot{q}_s + N_x(q,\dot{q}) = F_x(q)\tau \tag{4.18}$$

$$m_{sx}(q_s)\ddot{q}_x + m_{ss}(q_s)\ddot{q}_s + N_s(q,\dot{q}) = F_s(q)\tau \tag{4.19}$$

where $q = (q_x, q_s) \in Q = Q_x \times Q_s$, $\tau \in \mathbb{R}^m$, $F(q) = \mathrm{col}(F_x(q), F_s(q))$, and $\mathrm{rank}(F(q)) < n = \dim(q)$.

In his analysis and synthesis, *Olfati-Saber* has considered a certain number of cases based on complete, partial or non-actuation of the shape variables, of inputs coupling and of generalized momentums integrability. These properties and others are summarized as follows:

- When the shape variables are actuated for non-coupled inputs, this corresponds to the situation where $F_x(q) = 0$ and $F_s(q) = I_m$.
- When the shape variables are non-actuated for non-coupled inputs, this corresponds to the situation where $F_x(q) = I_m$ and $F_s(q) = 0$.
- When the inputs are coupled this corresponds, without loss of generality, to the situation where $F_x(q) \neq 0$ and $F_s(q)$ is a $m \times m$ matrix.
- When the inertia matrix is constant then the associated system is said to be flat.
- When the normalized generalized momentum defined by

$$\pi_x = \dot{q}_x + m_{xx}^{-1}(q_s)m_{xs}(q_s)\dot{q}_s \quad \text{or by} \quad \pi_s = \dot{q}_x + m_{sx}^{-1}(q_s)m_{ss}(q_s)\dot{q}_s$$

conjugated to q_x or q_s, respectively, is said to be integrable, then there exists a function $h = h(q_x, q_s)$ such that $\dot{h} = \pi_x$ or (π_s); otherwise the momentum is said to be non-integrable. In this case, the procedure is to decompose the non-integrable momentum into two terms. An integrable momentum called 'locked momentum' and a non-integrable momentum called 'error momentum'. For example: $\pi_x = \pi_x' + \pi_x^e$. Additionally, π^e is not dependent on $(q_x, \dot{q}_x)$ and becomes 0 when $q_s = \bar{q}_s$, the variable of the locked form.

All in all, this procedure has led *Olfati-Saber* to establish 16 classes for UMSs. Nevertheless, due to the redundancy of certain classes and the fact that some of them are physically non-realizable, the author reduced the 16 classes to only eight different classes.

For each of them, *Olfati-Saber* has proposed an order reduction method and a global change of coordinates allowing to transform these systems into three normal cascaded forms: strict feedback normal form, feedforward normal form, and non-triangular quadratic form as defined previously.

In summary, this classification, based on the structural properties of the UMSs—such as the actuation or non-actuation of the shape variables, coupling or non-coupling of the inputs, the integrability or non-integrability of the generalized momentum and supplementary condition—is illustrated in Fig. 4.8.

Remark 4.8

1. First, note that some examples appear in several classes. In fact, for the same system, we can have underactuation due to the absence of motors at different locations as is the case for the Acrobot and the Pendubot.
2. Next, the change of coordinates and normal forms obtained for the Classes I, IIa, IIb, and III are the same for the UMSs with two degrees of freedom. It suffices to replace $(\cdot)_1$ by $(\cdot)_x$ and $(\cdot)_2$ by $(\cdot)_s$.
3. Finally, for the other classes the change of coordinates and the obtained normal forms are given in [4].

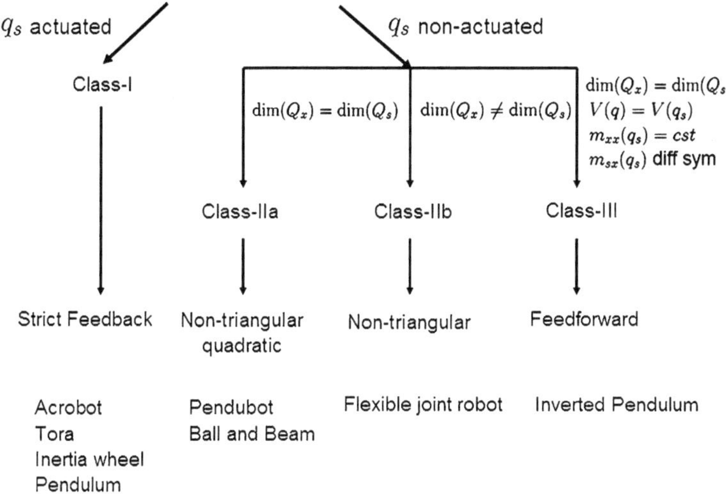

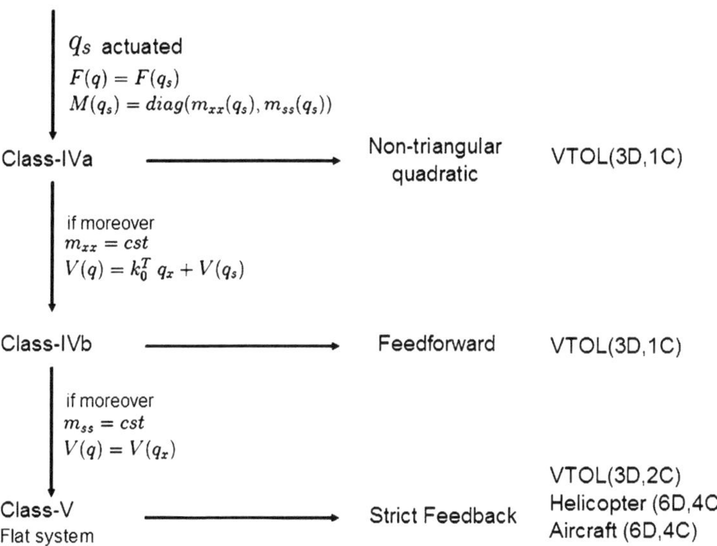

Fig. 4.8 Classification of UMSs according to *Olfati-Saber*

4.3 Comparison Between the Classifications

One of the objectives of this chapter is to show whether or not there exist some links between the available classifications in the literature.

First of all, we can obviously see that the two classification approaches are different, the first classification is graphical while the second one is rather analytical.

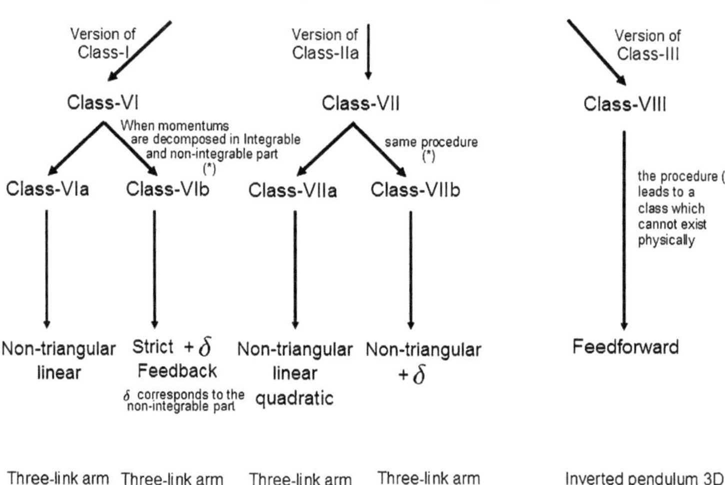

Fig. 4.8 (Continued)

In fact, the *Seto and Baillieul* classification is based on the construction of the control diagram flow, which reflects the way generalized forces are transmitted through components and highlights the interaction between DOF. For UMSs with 2 DOF, the procedure leads to the definition of three main classes according to the obtained structure in the CFD, namely: chain, tree, and isolated vertices structures. For higher order systems, the associated CFDs are combination of the three main structures, where seven classes are defined and classified according to the control degree of complexity. On the other hand, the *Olfati-Saber* classification is based on system structural properties such as kinetic symmetry, actuation of some variables, and integrability of generalized momentums. As a starting point, UMSs are transformed via partial linearizations into a cascade nonlinear and linear subsystems. However, the new control appears in both of the two subsystems. Then, explicit changes of coordinates are provided to decouple the two subsystems while leaving the linear subsystem invariant. The new cascade subsystems are in normal forms namely, systems in strict feedback form, feedforward form, and non-triangular linear-quadratic form. For UMSs with 2 DOF, three classes—Class I, Class II, and Class III—are defined according to the obtained normal form. For higher order systems, the nonlinear subsystems are reduced leading to eight classes.

Next, by constructing the CFDs for systems that belong to the same class in the classification of *Olfati-Saber*, we can readily notice that the obtained CFDs can have different structures see Fig. 4.9(a). In fact, the Acrobot and the Pendubot systems possess the same tree structure but belong to Classes I and II. Conversely, if we consider two systems having the same CFD structure in the *Seto and Baillieul* classification, then there is a possibility that they belong to two different classes in the

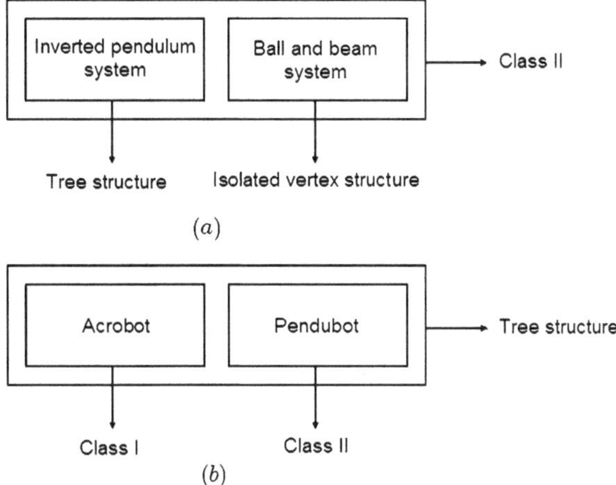

Fig. 4.9 Comparison between the two classifications

Olfati-Saber classification Fig. 4.9(b). Indeed, the ball and beam and the cart pole systems belong to Class II but their structures are different.

Finally, and from the control point of view, the authors of the first classification proposed a backstepping systematic control scheme applied in one step for a chain structure which is the less complicated structure to control. The systematic control for the other two structures is still an open problem. While the author of the second classification proposed a control algorithm for systems that can be transformed into a strict feedback normal form as well as for systems that can be transformed into the feedforward normal form. Some suggestions of control design for systems in non-triangular normal forms are also given. In each case, the control is built in two steps: First, the reduced model of the initial system is stabilized and then the global system is stabilized by a backstepping or a forwarding procedure depending on the associated normal form. *Olfati-Saber* has also given a theorem for the stabilization of the reduced systems associated to the strict feedback normal form. One of the hypothesis of the theorem is so restrictive that it is very rarely used and as a result, the synthesis is not systematic. Moreover, the control is built in two steps leading to complicated expressions.

It can be concluded that the two classifications are different and independent. In fact, the principle of classification and the control strategy are different, moreover, systems that belong to the same class in one classification belong to two different classes in the other one.

4.4 Summary

In this chapter, we have studied two approaches for the classification of UMSs. The first classification, due to *Seto and Baillieul,* is based on a control flow diagram that is constructed to represent the interaction forces through degrees of freedom of the underactuated system. Three structures are identified: chain, tree, and isolated vertices structures. From the combination of these structures one obtains seven structures. The second classification is due to *Olfati-Saber*, which takes into account the structural properties of the considered systems. It gives rise to eight classes that are transformable into three normal forms.

While attempting to find common points between these two classifications, it appeared that they are independent. Indeed, the ball and beam and the cart-inverted pendulum do not belong to the same class according to the first classification (isolated vertex structure and tree structure, respectively). However, they are in the same class (class II) for the other classification. Conversely, two systems of the same class in the *Seto* and *Baillieul* classification, for example the Acrobot and Pendubot, belong to two different classes in the *Olfati-Saber* classification, namely Class I and Class II. Moreover, classification procedures and control strategies are different.

References

1. A.M. Bloch, M. Reyhanoglu, N.H. McClamroch, Control and stabilization of nonholonomic dynamic systems. IEEE Trans. Autom. Control **37**(11), 1746–1757 (1992)
2. A.D. Luca, Dynamic control of robots with joint elasticity, in *Proc. 31st IEEE Conf. on Robotics and Automation* (1988), pp. 152–158
3. R. Olfati-Saber, Nonlinear control of underactuated mechanical systems with application to robotics and aerospace vehicles. Ph.D. thesis, Massachusetts Institute of Technology, Department Electrical Engineering and Computer Science, 2001
4. S.A. Raka, Systèmes mécaniques sous actionnés: classification, propriétés et contrôle. Master at, Aboubekr Belkaid University, Tlemcen Algeria, 2007
5. D. Seto, J. Baillieul, Control problems in super articulated mechanical systems. IEEE Trans. Autom. Control **39**(12), 2442–2453 (1994)

Chapter 5
Control Design Schemes for Underactuated Mechanical Systems

> *…A control theorist's first instinct in the face of a new problem is to find a way to use the tools he knows, rather than a commitment to understand the underlying phenomenon. This is not the failure of individuals but the failure of our profession to foster the development of experimental control science. In a way, we have become the prisoners of our rich inheritance and past successes.*
>
> *Y.C. Ho (1982)*

In this chapter, the stabilization issue for UMSs is considered. The strategy employed is based on the classification of *Seto and Baillieul* for these systems. The authors of this classification proposed a systematic control design procedure of backstepping type for the chain structure only. We are therefore concerned here with the problem of synthesizing control laws for each of the structures of this classification; thus, providing a general treatment of all the UMSs. For this, we shall firstly extend the procedure of *Seto and Baillieul* to a subclass of the tree structure that can be transformed to a chain structure under some conditions. Next, a procedure to control the remaining tree structure that cannot be transformed into a chain structure is presented. Finally, the control of the isolated vertex structure, which is the most difficult structure to control, is proposed.

5.1 Stabilization of Underactuated Systems in Chained Form

In this section, we recall the systematic backstepping procedure due to *Seto and Baillieul* [18], for the stabilization of UMSs with chain structure. Additionally, we establish different proofs associated with this procedure and illustrate the latter through the sliding-mass system example.

In a chain structure, the configuration variables and the controls are in series and each configuration variable (DOF) belongs to a single control path.

We consider the case where the CFD, contains one single control path. Whenever this not the case, each control path will be treated independently.

A. Choukchou-Braham et al., *Analysis and Control of Underactuated Mechanical Systems*, DOI 10.1007/978-3-319-02636-7_5,

The most general representation of this structure is given by the following triangular form:

$$\ddot{x}_i = N_i(x_1, \ldots, x_{i+1}, \dot{x}_1, \ldots, \dot{x}_{i+1}), \quad i = 1, \ldots, n-1$$
$$\ddot{x}_n = N_n(x, \dot{x}) + G(x, \dot{x})u$$

$$(5.1)$$

where $x = [x_1, \ldots, x_n]^T \in \mathbb{R}^n$ and $N_i(\cdot)$, $G(\cdot)$ are smooth functions.

The main idea here is to apply a method inspired by feedback linearization without explicitly linearizing the system. In this case, the coordinates transformation is avoided, and the variables keep their physical meaning.

However, this approach requires the following hypotheses:

(H1) $N_i(0) = 0$, $i = 1, \ldots, n$, the origin is an equilibrium.

(H2) For each $i = 1, \ldots, n-1$, $N_i(\cdot)$ are smooth functions with bounded states $x_1, \ldots, x_i$; $\dot{x}_1, \ldots, \dot{x}_i$, N_i is bounded only if x_{i+1} and $\dot{x}_{i+1}$ are bounded.

    In the case of feedback linearization, the latter hypothesis is equivalent to saying that the nonlinear subsystem of the normal form has the property of being BIBS (Bounded Input Bounded State), which is a necessary condition to avoid the peaking phenomenon in global stabilization.

(H3) $G(x, \dot{x}) \neq 0$ and either $\begin{cases} \partial N_i/\partial \dot{x}_{i+1} \neq 0 \text{ or} \\ \partial N_i/\partial \dot{x}_{i+1} = 0 \text{ but } \partial N_i/\partial x_{i+1} \neq 0 \end{cases}$

    This hypothesis ensures that the system is controllable, and that the different DOF of the chain are linked to each other.

(H4) For all $\partial N_i/\partial \dot{x}_{i+1} \neq 0$, $i = 1, \ldots, n-1$, the nonlinear system $N_i(0, \ldots, x_{i+1}, 0, \ldots, \dot{x}_{i+1}) = 0$ is GAS at $x = 0$ or when $\partial N_i/\partial \dot{x}_{i+1} = 0$ but $\partial N_i/\partial x_{i+1} \neq 0$ then the nonlinear system $N_i(0, \ldots, x_{i+1}, 0, \ldots, 0) = 0$ is GAS at $x = 0$.

    This hypothesis is equivalent to the global stability of the zero dynamics.

Remark 5.1 We assume that $y = x_1$. This choice is justified by the fact that we want to control the degree of freedom that is farthest from the input; that is, x_1.

To describe the stabilization results, we first define the following sequences.

Let us define $\bar{x}_1 = \begin{bmatrix} x_1 \\ \dot{x}_1 \end{bmatrix}$, $b = \begin{bmatrix} 0 \\ 1 \end{bmatrix}$, $e_1 = \bar{x}_1^T P b$ (P a positive definite matrix with all positive elements), $G_1 = 1$ and $W_1 = 0$.

For $i = 1, \ldots, n-1$

$$\left. \begin{aligned} e_{i+1} &= G_i N_i + W_i + k_i e_i, \\ G_{i+1} &= \frac{\partial N_i}{\partial \dot{x}_{i+1}} G_i, \\ W_{i+1} &= \sum_{j=1}^{i+1} \frac{\partial e_{i+1}}{\partial x_j} \dot{x}_j + \sum_{j=1}^{i} \frac{\partial e_{i+1}}{\partial \dot{x}_j} N_j + e_i, \end{aligned} \right\} \quad \text{if } \frac{\partial N_i}{\partial \dot{x}_{i+1}} \neq 0 \qquad (5.2)$$

$$\left.\begin{aligned}
e_{i+1} &= G_{i+1}\dot{x}_{i+1} + W_{(i+1)1} + k_{(i+1)1}e_{(i+1)1}, \\
e_{(i+1)1} &= G_i N_i + W_i + k_i e_i, \\
G_{i+1} &= \frac{\partial N_i}{\partial x_{i+1}} G_i, \\
W_{i+1} &= \sum_{j=1}^{i+1} \frac{\partial e_{i+1}}{\partial x_j}\dot{x}_j + \sum_{j=1}^{i} \frac{\partial e_{i+1}}{\partial \dot{x}_j} N_j + e_{(i+1)1}, \\
W_{(i+1)1} &= \sum_{j=1}^{i}\left(\frac{\partial e_{(i+1)1}}{\partial x_j}\dot{x}_j + \frac{\partial e_{(i+1)1}}{\partial \dot{x}_j} N_j \right) + e_i,
\end{aligned}\right\} \quad \text{if } \frac{\partial N_i}{\partial \dot{x}_{i+1}} = 0 \qquad (5.3)$$

where $k_{(i+1)1}$, k_i, $i = 1, \ldots, n-1$, and k_n are positive constants.

Theorem 5.1 [18] *Under the hypotheses* H1–H4, *the system* (5.1) *is globally asymptotically stable at the origin if the control law is chosen as follows*:

$$u = -(G_n G)^{-1}(G_n N_n + W_n + k_n e_n) \qquad (5.4)$$

Remark 5.2

- There is an adaptive version of this theorem in the case of parameter uncertainties, see [19].
- There also exists a version of this theorem in the case of trajectory tracking.
- The authors in [19] gave a proof of Theorem 5.1 in the adaptive case and for a control issued from the first sequence (5.2). In the following, we shall give the proof of the Theorem 5.1 in the absence of uncertainties and for controls issued from the two sequences (5.2) and (5.3), since $\frac{\partial N_i}{\partial \dot{x}_{i+1}} = 0$ for some considered applications.

Proof When $\frac{\partial N_i}{\partial \dot{x}_{i+1}} \neq 0$, the control is calculated from the sequence (5.2).

For each degree of freedom, x_i, the variable x_{i+1} is the "control variable" that controls the behavior of x_i. We also calculate the reference velocity, $\dot{x}_{r_{i+1}}$ for $\dot{x}_{i+1}$ such that when $\dot{x}_{i+1} \rightarrow \dot{x}_{r_{i+1}}$, x_i behaves as desired.

Step 1, $i = 1$

$$\ddot{x}_1 = N_1(x_1, x_2, \dot{x}_1, \dot{x}_2) \qquad (5.5)$$

Let us define a reference velocity such that

$$\dot{x}_{r2} = \dot{x}_2 - N_1 - k_1 x_1 - k_2 \dot{x}_1$$

The error between the reference velocity and the actual velocity is given by

$$e_2 = \dot{x}_2 - \dot{x}_{r2} = N_1 + k_1 x_1 + k_2 \dot{x}_1 \quad \Rightarrow \quad N_1 = e_2 - k_1 x_1 - k_2 \dot{x}_1$$

Let

$$\bar{x}_1 = \begin{pmatrix} x_1 \\ \dot{x}_1 \end{pmatrix}, \qquad A = \begin{pmatrix} 0 & 1 \\ -k_1 & -k_2 \end{pmatrix}, \qquad b = \begin{pmatrix} 0 \\ 1 \end{pmatrix}$$

such that $\dot{\bar{x}}_1$ can be expressed by

$$\dot{\bar{x}}_1 = \begin{pmatrix} \dot{x}_1 \\ \ddot{x}_1 \end{pmatrix} = \begin{pmatrix} 0 & 1 \\ -k_1 & -k_2 \end{pmatrix} \begin{pmatrix} x_1 \\ \dot{x}_1 \end{pmatrix} + \begin{pmatrix} 0 \\ 1 \end{pmatrix} e_2$$

where k_1 and k_2 are chosen such that

$$\ddot{x}_1 + k_2 \dot{x}_1 + k_1 x_1 = 0$$

is asymptotically stable at $(x_1, \dot{x}_1) = (0, 0)$. This implies the existence of a definite positive matrix P such that

$$A^T P + PA = -Q < 0$$

By applying the above definitions to (5.5), we get

$$\dot{\bar{x}}_1 = A\bar{x}_1 + be_2$$

Consider the following scalar function:

$$V_1 = \frac{1}{2}\left(\bar{x}_1^T P \bar{x}_1 + e_2^2\right) \tag{5.6}$$

The time derivative $\dot{V}_1$ is given by

$$\dot{V}_1 = \frac{1}{2}\left(\dot{\bar{x}}_1 P \bar{x}_1 + \bar{x}_1^T P \dot{\bar{x}}_1\right) + e_2 \dot{e}_2$$

$$= \frac{1}{2}\left(\bar{x}_1^T A^T P \bar{x}_1 + \bar{x}_1^T P A \bar{x}_1 + (be_2)^T P \bar{x}_1 + \bar{x}_1^T P(be_2)\right) + e_2 \dot{e}_2$$

$$= -\frac{1}{2}\left(\bar{x}_1^T Q \bar{x}_1\right) + \bar{x}_1^T P be_2 + e_2 \dot{e}_2$$

If we set

$$e_1 = \bar{x}_1^T Pb$$

and

$$nu_1 = \frac{1}{2}\bar{x}_1^T Q \bar{x}_1$$

then

$$\dot{V}_1 = -\nu_1 + e_2(\dot{e}_2 + e_1)$$

$$= -\nu_1 + e_2(\dot{N}_1 + k_1 \dot{x}_1 + k_2 \ddot{x}_1 + e_1)$$

$$= -\nu_1 + e_2\left(\underbrace{\frac{\partial N_1}{\partial x_1}\dot{x}_1 + \frac{\partial N_1}{\partial x_2}\dot{x}_2 + \frac{\partial N_1}{\partial \dot{x}_1}\ddot{x}_1 + \underbrace{\frac{\partial N_1}{\partial \dot{x}_2}}_{\stackrel{\text{def}}{=} G_2} \underbrace{\ddot{x}_2}_{\overset{=}{N_2}} + k_1 \dot{x}_1 + k_2 \ddot{x}_1 + e_1}\right)$$

$$= -v_1 + e_2\left(G_2N_2 + \left(\frac{\partial e_2}{\partial x_1} - k_1\right)\dot{x}_1 + \frac{\partial e_2}{\partial x_2}\dot{x}_2 + \left(\frac{\partial e_2}{\partial \dot{x}_1} - k_2\right)\ddot{x}_1\right.$$

$$\left. + k_1\dot{x}_1 + k_2\ddot{x}_1 + e_1\right)$$

$$= -v_1 + e_2\Big(G_2N_2 + \underbrace{\frac{\partial e_2}{\partial \dot{x}_1}\dot{x}_1 + \frac{\partial e_2}{\partial x_2}\dot{x}_2 + \frac{\partial e_2}{\partial \dot{x}_1}N_1 + e_1}_{\overset{\text{def}}{=}W_2}\Big)$$

$$= -v_1 + e_2(G_2N_2 + W_2)$$

The new control $\dot{x}_3$ appears through N_2. Consequently, we can choose $\dot{x}_{r_3}$ such that $e_2(G_2N_2 + W_2)$ is non-positive.

Step 2, $i = 2$

$$\dot{x}_{r3} = \dot{x}_3 - G_2N_2 - W_2 - k_2e_2$$

The difference between the reference and the actual velocity is given by

$$e_3 = \dot{x}_3 - \dot{x}_{r3} = G_2N_2 + W_2 + k_2e_2 \quad \Rightarrow \quad G_2N_2 + W_2 = e_3 - k_2e_2$$

In this case

$$\dot{V}_1 = -v_1 + e_2(e_3 - k_2e_2)$$
$$= -\left(v_1 - k_2e_2^2\right) + e_2e_3$$
$$= -v_2 + e_2e_3$$

with $v_2 = v_1 + k_2e_2^2$.

To compensate for e_3, we modify the function V_1 as

$$V_2 = V_1 + \frac{1}{2}e_3^2$$

By differentiation of V_2, we get

$$\dot{V}_2 = \dot{V}_1 + e_3\dot{e}_3$$

By the same calculation, we obtain the expression of V_2:

$$\dot{V}_2 = -v_2 + e_3(G_3N_3 + W_3)$$

where e_2 is in the expression of W_3 and $G_3 = G_2\frac{\partial N_2}{\partial \dot{x}_3}$.

The new control $\dot{x}_4$ appears through N_3, consequently we can choose $\dot{x}_{r_4}$ such that $e_3(G_3N_3 + W_3)$ is non-positive.

$$\vdots$$

Step n

After $(n-1)$ iterations, we can write

$$\dot{V}_{n-1} = -v_{n-1} + e_n(G_n \ddot{x}_n + W_n)$$
$$= -v_{n-1} + e_n(G_n N_n + G_n G u + W_n)$$

with

$$v_{n-1} = \frac{1}{2}\bar{x}_1^T Q\bar{x}_1 + \sum_{i=2}^{n-1} k_i e_i^2$$

$$\ddot{x}_n = N_n + Gu$$

$$G_n = \prod_{j=1}^{n-1} \frac{\partial N_j}{\partial \dot{x}_{j+1}}$$

In order to have $\dot{V}_{n-1} < 0$, it is sufficient to choose a control input u such that

$$u = -(G_n G)^{-1}(G_n N_n + W_n + k_n e_n)$$

The Lyapunov function V that guarantees the global asymptotic stability is such that

$$V_n = V_{n-1}$$

Hence

$$\dot{V}_n = -v_{n-1} - k_n e_n^2$$

Note that $G_n G$ is invertible because $G_n G$ and G are different from zero by hypothesis ($G \neq 0$ to ensure the controllability and $G_n \neq 0$ as a consequence of hypothesis H3). $\qquad\square$

Proof When $\frac{\partial N_i}{\partial \dot{x}_{i+1}} = 0$, the control is calculated from the sequence (5.3).

In order to keep the calculations simple and because the most of the examples considered are systems with 2 DOF, we restrict ourselves to the case $n = 2$.

Step 1, i = 1

When $\frac{\partial N_1}{\partial \dot{x}_2} = 0$ and $\frac{\partial N_1}{\partial x_2} \neq 0$, the differential equation is reduced to

$$\ddot{x}_1 = N_1(x_1, x_2, \dot{x}_1) \tag{5.7}$$

In the same manner, we come back to the preceding proof for $\frac{\partial N_i}{\partial \dot{x}_{i+1}} = 0$. In this case, instead of looking for a reference velocity $\dot{x}_{r_2}$ for the variable $\dot{x}_2$, we begin by looking for a reference position x_{r_2}: $x_{r2} = x_2 - N_1 - k_1 x_1 - k_2 \dot{x}_1$.

The difference between the reference and the actual position is given by

$$e_{21} = x_2 - x_{r2} = N_1 + k_1 x_1 + k_2 \dot{x}_1 \quad \Rightarrow \quad N_1 = e_{21} - k_1 x_1 - k_2 \dot{x}_1$$

Let $\bar{x}_1$, A and b as defined previously. Using this notation, we obtain

$$\dot{\bar{x}}_1 = A\bar{x}_1 + be_{21}$$

Let the candidate Lyapunov function be given this time by

$$V_{11} = \frac{1}{2}\left(\bar{x}_1^T P\bar{x}_1 + e_{21}^2\right) \tag{5.8}$$

The derivative $\dot{V}_{11}$ yields

$$\dot{V}_{11} = -\frac{1}{2}\bar{x}_1^T Q\bar{x}_1 + \bar{x}_1^T Pbe_{21} + e_{21}\dot{e}_{21}$$

If $e_1 = \bar{x}_1^T Pb$ and $v_{11} = \frac{1}{2}\bar{x}_1^T Q\bar{x}_1$, then

$$
\begin{aligned}
\dot{V}_{11} &= -v_{11} + e_{21}(\dot{e}_{21} + e_1)\\
&= -v_{11} + e_{21}(\dot{N}_1 + k_1\dot{x}_1 + k_2\ddot{x}_1 + e_1)\\
&= -v_{11} + e_{21}\left(\frac{\partial N_1}{\partial x_1}\dot{x}_1 + \frac{\partial N_1}{\partial x_2}\dot{x}_2 + \frac{\partial N_1}{\partial \dot{x}_1}\ddot{x}_1 + \frac{\partial N_1}{\partial \dot{x}_2}\ddot{x}_2\right.\\
&\qquad\left. + k_1\dot{x}_1 + k_2\ddot{x}_1 + e_1\right)\\
&= -v_{11} + e_{21}\left(\left(\frac{\partial e_{21}}{\partial x_1} - k_1\right)\dot{x}_1 + \frac{\partial e_{21}}{\partial x_2}\dot{x}_2 + \left(\frac{\partial e_{21}}{\partial \dot{x}_1} - k_2\right)\ddot{x}_1\right.\\
&\qquad\left. + k_1\dot{x}_1 + k_2\ddot{x}_1 + e_1\right)\\
&= -v_{11} + e_{21}\left(\underbrace{\frac{\partial N_1}{\partial x_2}}_{\overset{\text{def}}{=}G_2}\dot{x}_2 + \underbrace{\frac{\partial e_{21}}{\partial x_1}\dot{x}_1 + \frac{\partial e_{21}}{\partial \dot{x}_1}N_1 + e_1}_{\overset{\text{def}}{=}W_{21}}\right)\\
&= -v_{11} + e_{21}(G_2\dot{x}_2 + W_{21})
\end{aligned}
$$

This time, we cannot reach u through $\dot{x}_2$ but through $\ddot{x}_2$; so we add a supplementary step to the preceding proof. This additional step consists in finding a reference velocity $\dot{x}_{r2}$ for $\dot{x}_2$ such that $e_{21}(G_2\dot{x}_2 + W_{21})$ is non-positive:

$$\dot{x}_{r2} = \dot{x}_2 - G_2\dot{x}_2 - W_{21} - k_{21}e_{21}$$

The difference between the reference and the actual velocity is

$$e_2 = \dot{x}_2 - \dot{x}_{r2} = G_2\dot{x}_2 + W_{21} + k_{21}e_{21} \quad \Rightarrow \quad G_2\dot{x}_2 + W_{21} = e_2 - k_{21}e_{21}$$

Hence

$$\dot{V}_{11} = -v_{11} + e_{21}(e_2 - k_{21}e_{21})$$

$$= -v_{11} - k_{21}e_{21}^2 + e_{21}e_2$$

$$= -v_1 + e_{21}e_2$$

with $v_1 = v_{11} + k_{21}e_{21}^2$.

In order to compensate for e_2, the function V_{11} is modified as

$$V_1 = V_{11} + \frac{1}{2}e_2^2$$

By differentiation of V_1, we obtain

$$\dot{V}_1 = \dot{V}_{11} + e_2\dot{e}_2$$

$$= -v_1 + e_{21}e_2 + e_2\dot{e}_2$$

$$= -v_1 + e_2(\dot{e}_2 + e_{21})$$

$$= -v_1 + e_2\left(\underbrace{\frac{\partial e_2}{\partial x_1}\dot{x}_1 + \frac{\partial e_2}{\partial x_2}\dot{x}_2 + \frac{\partial e_2}{\partial \dot{x}_1}\ddot{x}_1 + e_{21}}_{\overset{\text{def}}{=}W_2} + \underbrace{\frac{\partial e_2}{\partial \dot{x}_2}\ddot{x}_2}_{G_2}\right)$$

$$= -v_1 + e_2(G_2\ddot{x}_2 + W_2)$$

$$= -v_1 + e_2\big(G_2(N_2 + Gu) + W_2\big)$$

Finally, the time derivative of the Lyapunov function is given by

$$\dot{V}_1 = -v_1 + e_2(G_2N_2 + G_2Gu + W_2) \tag{5.9}$$

In order to make $\dot{V}_1$ non-positive, u can be chosen such that

$$e_2(G_2N_2 + G_2Gu + W_2) = -k_2e_2^2 \tag{5.10}$$

Hence, the expression of the control which globally asymptotically stabilize the system when $n = 2$ is given by

$$u = -(G_2G)^{-1}(G_2N_2 + W_2 + k_2e_2)$$

For the same reasons as before, G_2G is invertible, which guarantees the existence of the control law.

Step 2, i = 2

The final Lyapunov function is given by

$$V_2 = V_1$$

such that

$$\dot{V}_2 = -v_1 - k_2e_2^2 \qquad\qquad \square$$

Example 5.1 To illustrate this control procedure, we consider an UMS that naturally has a chain structure; the mass sliding on cart system represented by Fig. 3.3. The dynamical model of the system, as given in Chap. 3, is described by the following equations:

$$m\ddot{x}_1 - B(\dot{x}_1 - \dot{x}_2) = 0$$

$$M\ddot{x}_2 + B(\dot{x}_1 - \dot{x}_2) = \tau$$

where m, M are, respectively, the masses of the small mass and the cart, B is the friction (which can be nonlinear) between the two masses.

The triangular representation of this model is given by

$$\ddot{x}_1 = \frac{B}{m}(\dot{x}_1 - \dot{x}_2) = N_1(\dot{x}_1, \dot{x}_2)$$

$$\ddot{x}_2 = \frac{1}{M}\big[\tau - B(\dot{x}_1 - \dot{x}_2)\big] = N_2(\dot{x}_1, \dot{x}_2) + G\tau$$

The CFD associated to this system, which was built in Chap. 4, is of chained form.

Consequently, the systematic procedure to synthesize a globally asymptotically stabilizing control can be applied to the sliding mass on cart. Nevertheless, before applying the procedure, it is necessary to verify whether the hypotheses H1–H4 are satisfied.

Remark 5.3

- Even if the control scheme presented before is applicable to every system that can be put in the form (5.1), the systems considered in this book are essentially of mechanical nature. In this case, it is customary to suppose, without loss of generality, that the state variables are bounded and consequently the property of BIBS is often satisfied.
- Moreover, we suppose that there is no rapid dynamics nor backlash in the gears.

  Verification of the hypotheses

(H1) Clearly, $N_i(0, 0) = 0$ for $i = 1, 2$. Hence, the origin is an equilibrium.

(H2) N_1 is a smooth function, N_1 is bounded for x_1 and x_2 bounded.

(H3) $G(x, \dot{x}) = \frac{1}{M} \Rightarrow G(x, \dot{x}) \neq 0$ and $\frac{\partial N_1}{\partial \dot{x}_2} = -\frac{B}{m} \Rightarrow \frac{\partial N_1}{\partial \dot{x}_2} \neq 0 \; \forall (x, \dot{x}) \in \mathbb{R}^{2n}$.

(H4) $N_1(0, \dot{x}_2) = 0 \Rightarrow -\frac{B}{m}\dot{x}_2 = 0 \Rightarrow \dot{x}_2 = 0$. This implies that the nonlinear system $N_1(0, \dot{x}_2) = 0$ is GAS at $(x_1, x_2) = 0$.

The hypotheses H1–H4 being verified, we can calculate the control law τ that ensures the GAS of the system. To compute this control we will use the sequences (5.2) because $\frac{\partial N_1}{\partial \dot{x}_2} \neq 0$. The use of these sequences leads to the following control law:

$$\tau = c_1(c_2 x_1 + c_3 \dot{x}_1 + c_4 \dot{x}_2) \tag{5.11}$$

where the c_i are combinations of the constants defined in the sequence (5.2).

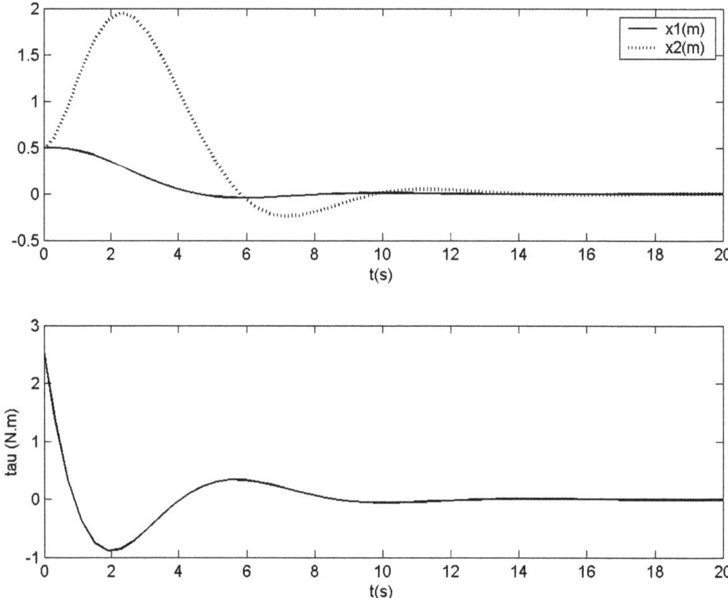

Fig. 5.1 State trajectories and the applied force to the mass sliding on the cart for the parameters $M = 1$ kg, $m = 0.2$ kg, $B = 0.02$ and for the initial conditions $(0.5, 0.5, 0, 0)$

Clearly, the obtained control law is simple and easy to implement. Note that for B constant, the system and the control law are linear. Hence, this procedure is also applicable to linear systems.

This control does not give any indication about the choice of the constants. However, they must be positive to ensure a negative derivative of the Lyapunov function.

The simulation of the controlled mass sliding on cart is illustrated in Fig. 5.1.

It can be seen that the motions of the mass and the cart stabilize rapidly at the origin with a small time response and a small control amplitude. Evidently, one can still modify the time response and the control effort by adjusting the constants.

Unfortunately, very few UMSs are naturally under the chained structure, the only examples found in the literature are those of the mass sliding on cart and the elastic joint robot presented in [6].

Effectively, most of the UMSs possess: either a tree structure, like the Acrobot, the Pendubot, the pendular systems, the inertia wheel pendulum and the Tora system, or an isolated vertex structure, like the ball and beam system (to cite only systems with two degrees of freedom).

As such there does not exist any systematic methodology to treat these two structures. Most of the time, they were dealt with on a case by case basis.

In what follows, we shall propose a systematic control design methodology for systems with tree structure. This will be done by transforming, under some conditions, a subclass of systems possessing a tree structure into that possessing a chain structure in such a way that the systematic procedure of backstepping presented earlier can be applied.

5.2 Systematic Control of Systems Possessing a Tree Structure

The construction of a CFD for a given system depends on the coordinates system and the external forces acting on it. Thus, the CFD is not invariant under a coordinate transformation. Owing to this simple remark it is important to find a change of coordinates that will transform the latter into a given form.

In fact, it will become apparent that if a given system satisfies the conditions enumerated in the paragraph below, a CFD that is under a tree structure can be transformed into that of chain structure. Whenever such is the case, one can use the same procedure proposed by *Seto and Baillieul* to synthesize a globally asymptotically stabilizing control. Evidently, we need to revert back to the initial system in order to retain the physical significance of the variables and the inputs.

Let us consider the general Euler–Lagrange equations of motion of an UMS:

$$m_{11}(q)\ddot{q}_1 + m_{12}(q)\ddot{q}_2 + h_1(q, \dot{q}) = \tau_1$$
$$m_{21}(q)\ddot{q}_1 + m_{22}(q)\ddot{q}_2 + h_2(q, \dot{q}) = \tau_2 \tag{5.12}$$

where $q \in Q$ is an n-dimensional manifold; we suppose that q can be written under the form $q = \mathrm{col}(q_1, q_2) \in Q_1 \times Q_2$ where Q_i is of dimension $n_i = \dim(Q_i)$ for $i = 1, 2$ and $n_1 + n_2 = n$.

The $m_{ij}(q)$ are the elements of $M(q)$, the inertia matrix of the system. The h_i's contain the Coriolis, centrifugal, and gravitational terms and the τ_i's are the controls satisfying one of the following two actuation conditions:

(A1) $\tau = \tau_2 \in \mathbb{R}^{n_2}$ is the control and $\tau_1 \equiv 0$.
(A2) $\tau = \tau_1 \in \mathbb{R}^{n_1}$ is the control and $\tau_2 \equiv 0$.

In the two cases, the system (5.12) is an underactuated one. The mode of actuation A1 or A2 is important and is defined depending on the applications. For example, the Acrobot, the Tora system, and the inertia wheel pendulum are actuated under the A1 mode, whereas for the Pendubot, the inverted pendulum is actuated under the A2 mode.

5.2.1 Stabilization of UMSs Actuated Under Mode A1

Let us make the following hypotheses:

(B1) The considered system possesses a kinetic symmetry; that is, $M(q) = M(q_2)$.
(B2) q_2 is the actuated variable; that is, the system is actuated under mode A1.
(B3) The quantity $m_{11}^{-1}(q_2)m_{12}(q_2)$ is integrable.

Note that the hypotheses are not very restrictive and are satisfied by a large class of UMSs.

We now present an important result in the following:

Theorem 5.2 *Assume that hypotheses* B1–B3 *are satisfied, then an underactuated system possessing a tree structure can be transformed into a system possessing a chain structure.*

Proof We have seen (Chap. 3, Sect. 3.7) that UMSs under mode A1 can be partially linearized using the following change of control input:

$$\tau = \alpha(q)u + \beta(q, \dot{q}) \tag{5.13}$$

which transforms the dynamics of (5.12) into

$$
\begin{aligned}
\dot{q}_1 &= p_1 \\
\dot{p}_1 &= f(q, p) + g_0(q)u \\
\dot{q}_2 &= p_2 \\
\dot{p}_2 &= u
\end{aligned}
\tag{5.14}
$$

where $\alpha(q)$ is an $m \times m$ SPD matrix and $g_0(q) = -m_{11}^{-1}(q)m_{12}(q)$.

Following this control input change, the new control appears both in the linear subsystem and the nonlinear one at the same time. So, we obtain a system with a tree CFD structure.

Also, *Olfati-Saber* showed that if an underactuated system satisfies hypotheses B1–B3, then it is transformable into a strict feedback normal form.

Effectively, the following change of coordinates:

$$
\begin{aligned}
q_r &= q_1 + \gamma(q_2) \\
p_r &= m_{11}(q_2)p_1 + m_{12}(q_2)p_2 := \frac{\partial L}{\partial \dot{q}_1}
\end{aligned}
\tag{5.15}
$$

transforms the dynamics of the system (5.14) into a nonlinear system in strict feedback cascade form:

$$
\begin{aligned}
\dot{q}_r &= m_{11}^{-1}(q_2)p_r \\
\dot{p}_r &= g(q_r, q_2) \\
\dot{q}_2 &= p_2 \\
\dot{p}_2 &= u
\end{aligned}
\tag{5.16}
$$

where

$$\gamma(q_2) = \int_0^{q_2} m_{11}^{-1}(\theta)m_{12}(\theta)\, d\theta$$

$$g(q_r, q_2) = -\frac{\partial V(q)}{\partial q_1}$$

Fig. 5.2 The Tora system

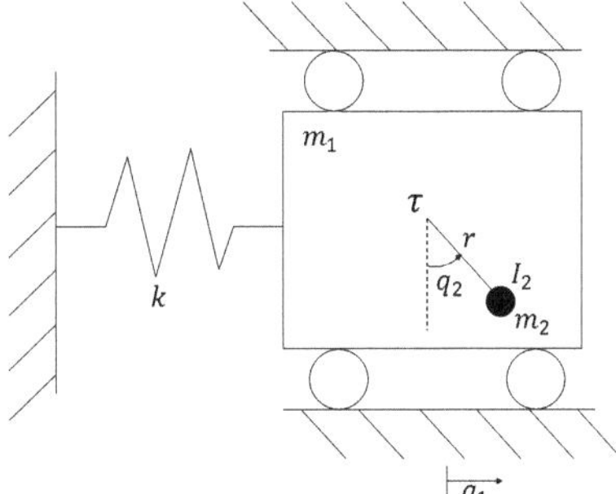

It suffices now to note that the form (5.16) can be put into a triangular one. More precisely, we can rewrite (5.16) in the form

$$\ddot{q}_r = m_{11}^{-1}(q_2)g(q_r, q_2)$$
$$\ddot{q}_2 = u$$
(5.17)

which is simply a form with a CFD in chain structure. Hence, the tree structure has been transformed into a chain structure.   □

Remark 5.4 For the A2 mode case, when q_2 is not actuated, there exist another change of control (non-collocated) and change of coordinates to transform the system into a normal form. However, the obtained normal form is not in strict feedback form. It means that some underactuated systems (especially those which are actuated under mode A2) cannot be transformed from a tree structure into a chain structure. This is the case of the Pendubot, the inverted pendulum, and some other systems. For such systems, we shall propose a strategy to construct stabilizing control laws in the following subsections.

In order to illustrate the procedure, let us consider a system that is initially in the tree structure, for example the Tora system.

5.2.1.1 Application: The Tora System

This system is composed of a platform that can oscillate on the horizontal plane without friction. On this platform, an eccentric mass is actuated by a DC motor. Its movement applies a force to the platform, which can be used to dampen the transverse oscillations; see Fig. 5.2.

The problem of stabilizing this system was introduced by Wan, Brenstein and Coppola [22] and has recently attracted the attention of many researchers by the fact that it presents a nonlinear interaction between the translational and rotational motions. It has also been used as a benchmark for the nonlinear control of cascaded systems, especially the passivity-based methods [12, 13, 17], backstepping [15, 22], robust control and sliding modes [10, 14, 24], dynamical surfaces [16],

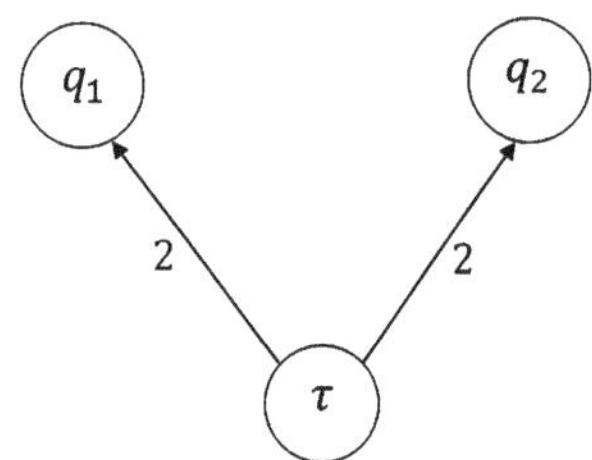

Fig. 5.3 CFD of the Tora system

LMI (linear matrix inequality) controllers [1, 9], velocity gradient [8], and fuzzy logic [11, 23]. Here, we shall present a switching control to stabilize the Tora system.

Suppose that the torque of the motor is the control variable. The objective is to determine a control law that stabilizes the rotational and translational motion at the origin, at the same time.

This system possesses two degrees of freedom (q_1, q_2), where q_1 is the unactuated variable and q_2 the actuated variable.

We recall the Euler–Lagrange equations describing the Tora system:

$$(m_1 + m_2)\ddot{q}_1 + m_2 r \cos(q_2)\ddot{q}_2 - m_2 r \sin(q_2)\dot{q}_2^2 + kq_1 = 0$$
$$m_2 r \cos(q_2)\ddot{q}_1 + \left(m_2 r^2 + I_2\right)\ddot{q}_2 + m_2 gr \sin(q_2) = \tau \tag{5.18}$$

or

$$\ddot{q}_1 = \frac{1}{\det M(q_2)}\left(-m_2 r \cos(q_2)\tau + gm_2^2 r_2^2 \cos(q_2) \sin(q_2)\right.$$
$$\left. - \left(m_2 r^2 + I_2\right)\left(kq_1 - m_2 r \sin(q_2)\dot{q}_2^2\right)\right)$$
$$\ddot{q}_2 = \frac{1}{\det M(q_2)}\left((m_1 + m_2)\tau - (m_1 + m_2)m_2 gr \sin(q_2)\right.$$
$$\left. + m_2 r \cos(q_2)\left(kq_1 - m_2 r \sin(q_2)\dot{q}_2^2\right)\right) \tag{5.19}$$

where $\det M(q_2) = (m_1 + m_2)(m_2 r^2 + I_2) - (m_2 r \cos(q_2))^2$.

The construction of the corresponding CFD to this system is given by Fig. 5.3, which is clearly a tree structure.

After a partial linearization by the change of control:

$$\tau = \alpha(q)u + \beta(q, \dot{q}) \tag{5.20}$$

with

$$\alpha(q_2) = \left(m_2 r^2 + I_2\right) - \frac{(m_2 r \cos(q_2))^2}{m_1 + m_2} \quad \forall q_2 \in [-\pi, \pi]$$

$$\beta(q, \dot{q}) = m_2 gr \sin(q_2) - \frac{m_2 r \cos(q_2)}{m_1 + m_2}\left(kq_1 - m_2 r \sin(q_2)\dot{q}_2^2\right)$$

The dynamics of the Tora becomes

$$\dot{q}_1 = p_1$$
$$\dot{p}_1 = f_0(q, p) + g_0(q)u$$
$$\dot{q}_2 = p_2$$
$$\dot{p}_2 = u$$

(5.21)

with

$$f_0(q, p) = \frac{(m_2 r \sin(q_2))p_2 - kq_1}{m_1 + m_2}$$

$$g_0 = \frac{m_2 r \cos(q_2)}{m_1 + m_2}$$

Note that $M(q) = M(q_2)$ and that the Tora is actuated under mode A1. Also, the function $\gamma(q_2)$ can be explicitly calculated as

$$\gamma(q_2) = \int_0^{q_2} \frac{m_2 r \cos(\theta)}{m_1 + m_2} \, d\theta = \frac{m_2 r \sin(q_2)}{m_1 + m_2}$$

As all the hypotheses B1–B3 are verified, the following global coordinate change:

$$q_r = q_1 + \frac{m_2 r \sin(q_2)}{m_1 + m_2}$$
$$p_r = (m_1 + m_2)p_1 + m_2 r \cos(q_2)p_2$$

(5.22)

transforms the dynamics of the Tora into the following strict feedback cascaded nonlinear form:

$$\dot{q}_r = \frac{1}{(m_1 + m_2)}p_r$$
$$\dot{p}_r = -kq_r + k\gamma(q_2)$$
$$\dot{q}_2 = p_2$$
$$\dot{p}_2 = u$$

(5.23)

This system can also be written in the form

$$\ddot{q}_r = -\frac{k}{m_1 + m_2}q_r + \frac{km_2 r}{(m_1 + m_2)^2}\sin(q_2)$$
$$\ddot{q}_2 = u$$

(5.24)

which corresponds to a system of the form (5.1).

In this case, the CFD associated to (5.24) is given by Fig. 5.4.

Hence, the control change (5.20) and the coordinates change (5.22) transform the tree structure of the Tora into a chain structure.

Fig. 5.4 CFD of the transformed Tora

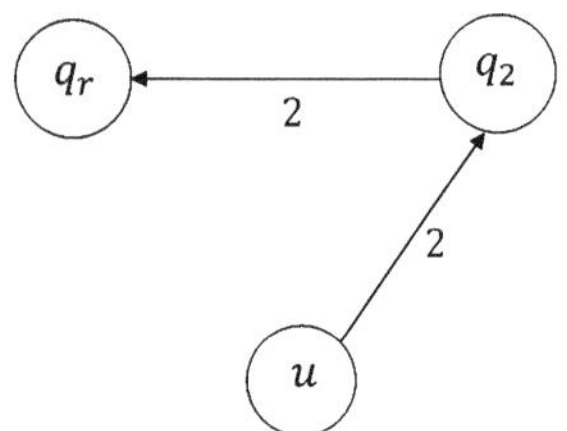

Now that the Tora is put into a chained structure, we can use the systematic procedure of *Seto and Baillieul* to derive a globally asymptotically stabilizing control law for the transformed Tora. Evidently, the real control to be applied to initial system is deduced by an inverse transformation, which is possible since the change of coordinates is global.

In order to be able to apply this procedure, let us first verify hypotheses H1–H4 for the system below:

$$\ddot{q}_r = -\frac{k}{m_1 + m_2}q_r + \frac{km_2r}{(m_1 + m_2)^2}\sin(q_2) = N_1(q_r, q_2)$$
$$\ddot{q}_2 = u = N_2 + Gu \quad \text{with } N_2 = 0 \text{ and } G = 1$$

(5.25)

As for the previous example, we suppose that there is no rapid dynamics nor backlash in the gears.

Verification of the hypotheses

(H1) $N_i(0, 0) = 0$, the origin is an equilibrium.
(H2) N_1 is a smooth function, N_1 is bounded for q_r and p_2 bounded.
(H3) $G(q_r, q_2) = 1 \Rightarrow G(q_r, q_2) \neq 0$.
   $\frac{\partial N_1}{\partial \dot{q}_2} = 0$ and $\frac{\partial N_1}{\partial q_2} = c \cos(q_2)$, ($c$ constant). Hence, $\frac{\partial N_1}{\partial q_2} \neq 0 \ \forall (q_r, q_2) \in D \subset \mathbb{R}^{2n}$ with $D = \{(q_r, q_2)/q_2 \neq (2k + 1)\pi/2\}$.
(H4) $N_1(0, q_2) = 0 \Rightarrow \sin(q_2) = 0 \Rightarrow q_2 = 0$. This implies that the nonlinear system $N_1(0, q_2) = 0$ is GAS at $(q_r, q_2) = 0$.

The application of the control scheme (5.3) leads to the following control law:

$$u_{nL} = -\frac{(m1 + m2)^2}{k\cos(q_2)}\left(c_1\dot{q}_r + \frac{k}{(m1 + m2)^2}\dot{q}_2\big(c_2\cos(q_2) - \dot{q}_2\big) + c_3 q_r + c_4\sin(q_2)\right)$$

(5.26)

where c_1, c_2, c_3, c_4 are positive constants.

Clearly, the obtained control is simple and easy to implement. Moreover, the rate of convergence can be controlled by adjusting the gains c_i.

Nevertheless, this control is valid only if $q_2 \neq (2k + 1)\pi/2$. This is because the hypothesis H3 is not always verified $\forall (q, \dot{q}) \in \mathbb{R}^{2n}$, as $\frac{\partial N_i}{\partial q_{i+1}} \neq 0$ only for $q_2 \neq (2k + 1)\pi/2$.

This means the existence of singularities in the control reducing the basin of attraction, which cannot be the entire space. Consequently, the stability cannot be global.

One solution to avoid the divergence of the state is to adjust the gains so to maintain the solution near the equilibrium. However, maintaining the solution near the equilibrium would imply slow convergence rate. Additionally, if the initial conditions for q_2 are chosen greater or equal to $\pi/2$ then the state will diverge due to this singularity. Hence, such solution must be avoided.

To solve this problem, we propose two solutions, the first one is based on switching control [4], the second one is based on Lyapunov method [3]. The two solutions will permit to find control laws that are valid for any initial conditions despite the singularity. As a result, the stability will be global.

Bypassing the Singularity: First Solution   The idea is to use an hybrid control that switches between the determined control law (5.26) far away from the singularities and another one near the singularities.

This control technique has been employed recently (Appendix A). Its importance comes from the fact that some systems cannot attain some objectives through a single control.

Let us determine the second control law. The idea is to use the linearized model of the Tora around the singularity to calculate a linear control to be applied in the neighborhood of this point. When the trajectories leave this neighborhood, we switch to the nonlinear control to achieve the global asymptotic stability of all the states.

The Expression of the Linear Control   The linearized model of the Tora system around $(q_r, p_r, q_2, p_2) = (0, 0, \pi/2, 0)$ is given by

$$
\begin{aligned}
\delta \dot{q}_r &= \frac{1}{(m_1 + m_2)} \delta p_r \\
\delta \dot{p}_r &= -k \delta q_r \\
\delta \dot{q}_2 &= \delta p_2 \\
\delta \dot{p}_2 &= \delta u
\end{aligned}
\tag{5.27}
$$

The new problem that appears is that the subsystem $(\delta q_r, \delta p_r)$ is uncontrollable. Fortunately, it is stable. According to Brockett in [2] (Theorem 3.1), when the uncontrollable modes are stable then the entire system can still be stabilized.

In this case, the expression of the linear control is determined in a standard manner and is given by

$$
u_L = -K * x
\tag{5.28}
$$

where $x = [\delta q_2, \delta p_2]^T$ and $K = [K_1 \; K_2]$ is a gain matrix calculated by either the LQR or by pole placement techniques (Appendix D).

The simulation of this hybrid control applied to the Tora system with the parameters $m_1 = 10$ kg, $m_2 = 1$ kg, $k = 5$ N/m, $r = 1$ m, $I = 1$ kg/m shows the effectiveness of the proposed control, see Fig. 5.5.

The control law stabilized the Tora system for critical initial conditions such that the singularity points towards itself, $q_2 = \pi/2$ (Fig. 5.6), or at a point further away from the singularity, $q_2 = \pi$ (Fig. 5.7).

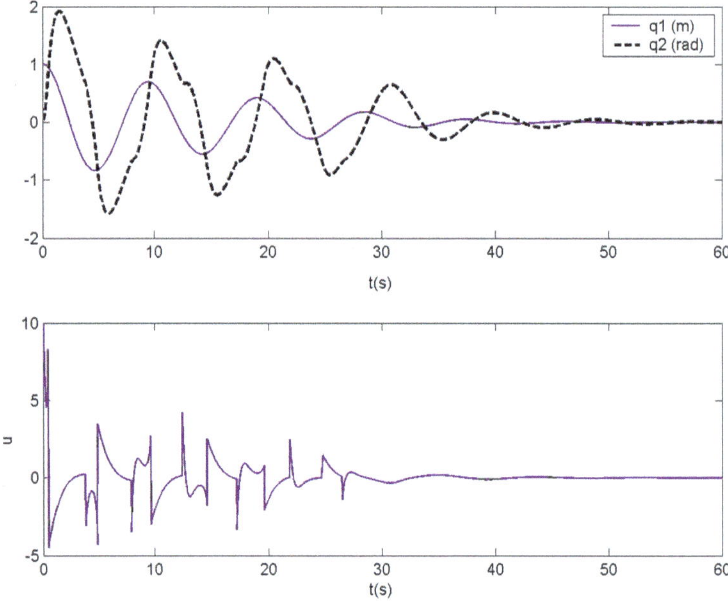

Fig. 5.5 States trajectories and input of the Tora system for the initial conditions $(q_1, q_2, p_1, p_2) = (1, 0, 0, 0)$

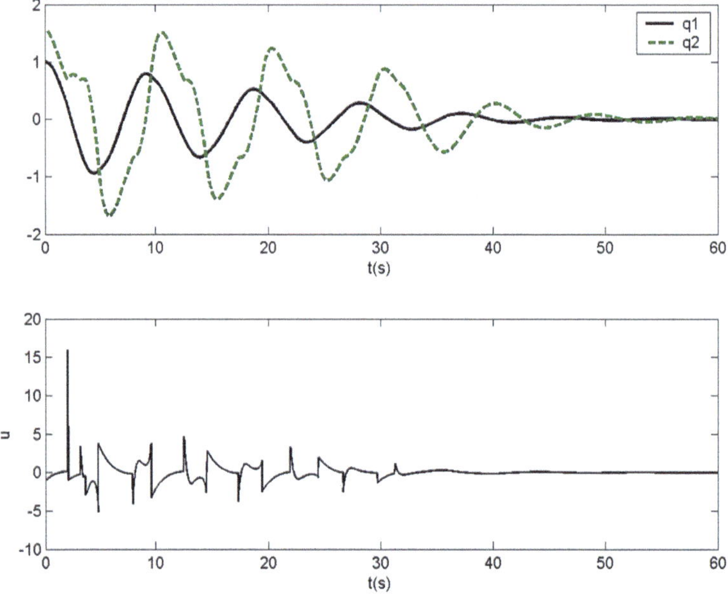

Fig. 5.6 States trajectories and input of the Tora system for the initial conditions $(q_1, q_2, p_1, p_2) = (1, \frac{\pi}{2}, 0, 0)$

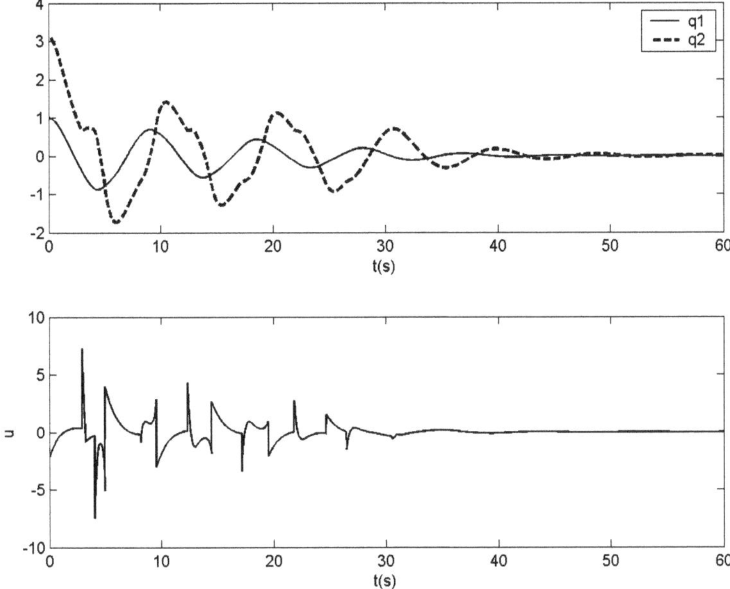

Fig. 5.7 States trajectories and input of the Tora system for the initial conditions $(q_1, q_2, p_1, p_2) = (1, \pi, 0, 0)$

Fig. 5.8 Switching intervals
for the control

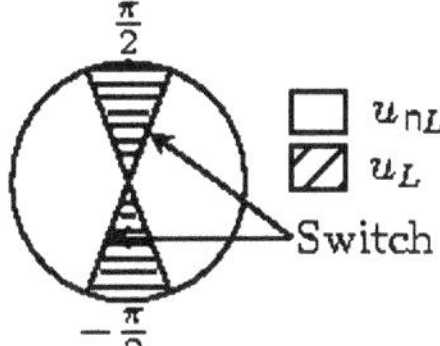

The switching from one control to the other is controlled by the state q_2, in such a way that when $|q_2|$ is outside the interval $\frac{\pi}{2} \pm e$, we apply the nonlinear control u_{nL} (5.26) and when $|q_2|$ enters this interval we switch to the linear control u_L (5.28), see Fig. 5.8.

The length of the interval is in direct relationship with the control effort. Effectively, one can notice that the small length of the interval corresponding to small values of e (around 0.2 to 0.3, Fig. 5.9) leads to much more important efforts than those corresponding to larger values of e (such that 0.5 to 0.6, Fig. 5.5). This is because with larger intervals we do not allow $\cos(q_2)$ to become very small in order to avoid large values for u_{nL}.

Stability Proof for the Hybrid Control   In Appendix A, we present the conditions that switching systems must satisfy in order to have a global stability for hybrid systems. In fact, we need either a common Lyapunov function to all subsystems

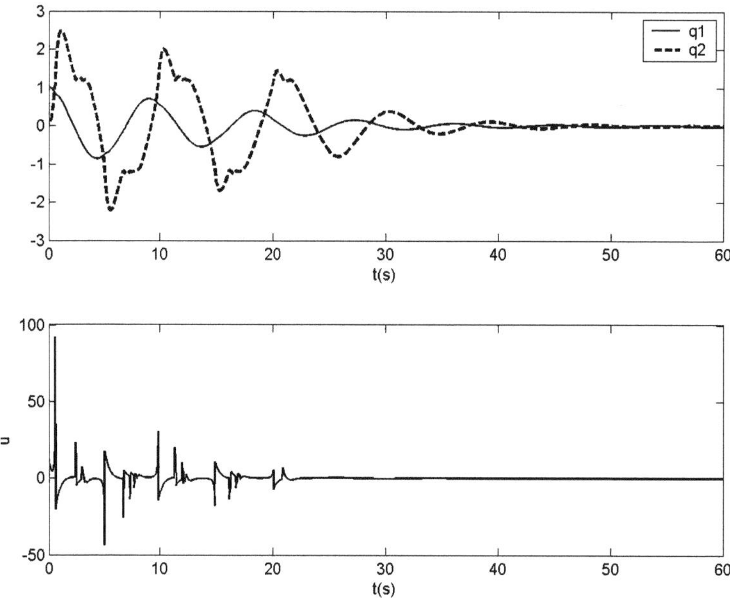

Fig. 5.9 State trajectories and the control of the Tora system for the initial conditions $(q_1, q_2, p_1, p_2) = (1, 0, 0, 0)$ and $e = 0.2$

or multiple Lyapunov functions, each one associated to one subsystem and some conditions on the values of the different Lyapunov functions at the switching times.

In the present case, we will prove the stability of the switching control system with multiple Lyapunov functions.

For the Tora system, these Lyapunov functions are:

$$V_{nL} = \frac{1}{2}\bar{q}_1^T P \bar{q}_1 + \frac{1}{2}e_{21}^2 + \frac{1}{2}e_2^2 \quad \text{for the nonlinear subsystem}$$

$$V_L = \frac{1}{2}\tilde{x}^T R \tilde{x} \qquad\qquad \text{for the linearized subsystem}$$

where $\bar{q}_1$, P, e_{21}, and e_2 are variables defined in the control sequences (5.3) for $\bar{q}_1 = \bar{x}_1$, $\tilde{x} = (\delta q_r, \delta p_r, \delta q_2, \delta p_2)$ are the linearized system's coordinates and R is a SPD matrix.

The proof of the nonlinear system stability under the action of the control u_{nL} with the first Lyapunov function has already been done earlier (Sect. 5.1). For the second subsystem (5.27), as it is linear we can choose a Lyapunov function of the form

$$V_L = \frac{1}{2}\tilde{x}^T R \tilde{x}$$

If the matrix R is chosen to be diagonal, then V_L can be expressed as

$$V_L = \frac{1}{2}\left(R_1\tilde{x}_1^2 + R_2\tilde{x}_2^2 + R_3\tilde{x}_3^2 + R_4\tilde{x}_4^2\right)$$

by differentiation of V_L, we obtain

$$\dot{V}_L = R_1\tilde{x}_1\dot{\tilde{x}}_1 + R_2\tilde{x}_2\dot{\tilde{x}}_2 + R_3\tilde{x}_3\dot{\tilde{x}}_3 + R_4\tilde{x}_4\dot{\tilde{x}}_4$$

$$= \left(\frac{R_1}{m_1 + m_2} - R_2k\right)\tilde{x}_1\tilde{x}_2 + (R_3 - K_1R_4)\tilde{x}_3\tilde{x}_4 - K_2R_4\tilde{x}_4^2$$

If the elements of the matrix R are chosen such that the conditions

$$\begin{cases} \dfrac{R_1}{m_1 + m_2} = R_2k \\ R_3 = K_1 \end{cases}$$

are satisfied, then

$$\dot{V}_L = -K_2R_4\tilde{x}_4^2$$

The linear subsystem is only stable, but the principle of invariance of Lasalle ensures the asymptotic stability.

Hence, the Lyapunov functions V_{nL} and V_L guarantee the global asymptotic stability of the nonlinear and linear subsystem, respectively. We have to show that these functions are decreasing on each interval where the subsystem is active. In other words, the value of the Lyapunov function at the end of an interval must be higher than its value at the end of next interval where the considered subsystem is active.

Unfortunately, it is very difficult to verify such conditions analytically, in most cases we resort to simulation [7]. In this case, we build by simulation the energy profile. That of the Tora system subjected to a switched control law is given in Fig. 5.10.

Based on this figure, the Lyapunov functions V_{nL} and V_L, satisfy the decreasing condition on the corresponding intervals. Consequently, the Tora system controlled by a switching control is globally asymptotically stable.

Bypassing the Singularity, Second Solution   The main purpose of this second solution is to avoid the use of a switching control. The idea is to determine a control which forces the derivative of the Lyapunov function (5.9) to be non-positive. The control (5.4) proposed by *Seto and Baillieul* is the most simple way to impose this condition. However, one needs to have the derivative of connection terms to be non-zero. This last condition is not necessarily satisfied by many underactuated systems, in particular, the Tora. We can choose another control which can force the derivative of the Lyapunov function (5.9) to be non-positive.

Let us therefore recall the expression of the derivative of the Lyapunov function for $N_2 = 0$ and $G = 1$:

$$\dot{V}_1 = -v_1 + e_2(G_2N_2 + G_2Gu + W_2)$$

$$= -v_1 + e_2G_2u + e_2W_2.$$

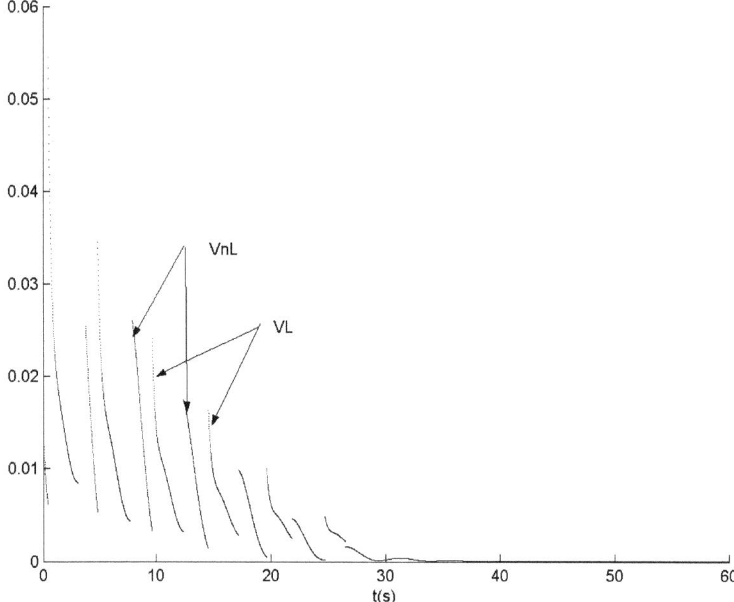

Fig. 5.10 Switching system energy profile

To force this expression to be negative, let us consider the following control law:

$$u = -ck_2 G_2 e_2 - c_{\text{var}} G_2 W_2 \tag{5.29}$$

where c is a positive constant used to adjust the rate of convergence associated to different initial conditions, while c_{var} is variable and will be defined later.

In this case

$$\dot{V}_1 = -v_1 - ck_2 G_2^2 e_2^2 - c_{\text{var}} G_2^2 e_2 W_2 + e_2 W_2$$

$$= -v_2 + e_2 W_2 \left(1 - c_{\text{var}} G_2^2\right)$$

with $v_2 = v_1 + ck_2 G_2^2 e_2^2$.

Now, we try to force the term $e_2 W_2 (1 - c_{\text{var}} G_2^2)$ to 0 such that $\dot{V}_1 = -v_2$.

To do that, we can choose c_{var} such that $c_{\text{var}} = \frac{1}{G_2^2}$.

Again, we find a division by zero. However, this time, the problem can be solved by modifying c_{var} into $c_{\text{var}} = \frac{1}{G_2^2 + E}$ near the singularities points, where

$$\begin{cases} E = 1 & \text{if } G_2 \approx 0 \text{ i.e. } |q_2| = |\pi/2 \pm 0.01| \\ E = 0 & \text{if } G_2 \neq 0 \text{ i.e. } |q_2| \neq |\pi/2 \pm 0.01| \end{cases}$$

Let us consider again the derivative of the Lyapunov function,

$$\dot{V}_1 = -v_2 + e_2 W_2 \left(1 - c_{\text{var}} G_2^2\right)$$

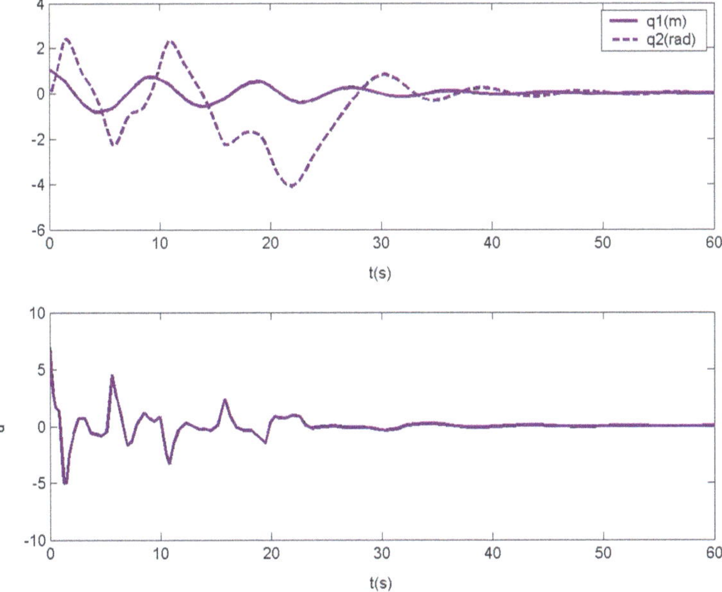

Fig. 5.11 States trajectories and control of the Tora system for the initial conditions $(q_1, q_2, \dot{q}_1, \dot{q}_2) = (1, 0, 0, 0)$

When the term $e_2 W_2(1 - c_{var} G_2^2) = 0$, then $\dot{V}_1 = -v_2$ and the global asymptotic stability is ensured, but when $G_2 = 0$, this term is reduced to $e_2 W_2$ such that $\dot{V}_1 = -v_2 + e_2 W_2$. To force the non-positiveness of the derivative of the Lyapunov function and since e_2 and W_2 according to (5.3) both depend on the constants k_1, k_2, k_{21}, and P_i (elements of P), we can then choose these constants such that the condition

$$|e_2 W_2| < |v_2| \tag{5.30}$$

is satisfied at all time.

The constants which force the condition (5.30) to hold true are given by $k_1 = 4$; $k_2 = 4$; $k_{21} = 0.08$; $p_2 = 0.1$; $p_4 = 0.1$. The simulations are done using the same Tora parameters as previously. The application of the proposed control (5.29), for the initial conditions $(q_1, q_2, \dot{q}_1, \dot{q}_2) = (1, 0, 0, 0)$ and for $c = 486$ gives the simulation results presented in Fig. 5.11.

It can be seen from the figure that for the controlled system, the two translational and rotational motions are stabilized at the origin with reasonable control effort and settling time.

The numerical verification of the condition (5.30) shows that the latter is always true Fig. 5.12.

It is important to note that the computed control law is still valid for any initial condition. Moreover, the rate of convergence can be adjusted with the constant c. To confirm this result we choose the initial condition $(q_1, q_2, \dot{q}_1, \dot{q}_2) = (1, \pi, 0, 0)$.

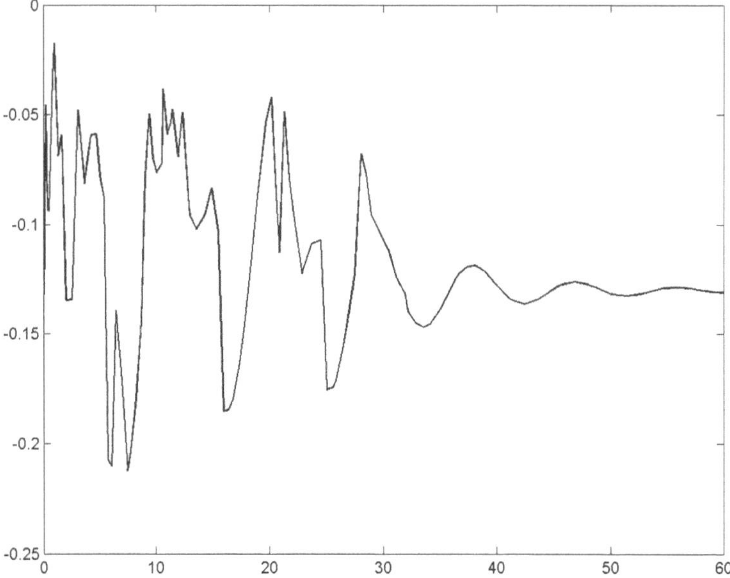

Fig. 5.12 Evolution of $|e_2 W_2| - |v_2|$

To stabilize this system from this initial condition, the trajectories will necessarily move through the singularities points. The corresponding simulation results, as depicted in Fig. 5.13 for $c = 500$, are very satisfying. It shows the effectiveness of the proposed control law in taking into account these singularities.

Let us consider now the stabilizing problem of the subclass of UMSs in a tree structure that cannot be transformed in a chain structure.

5.2.2 Stabilization of UMSs Actuated Under Mode A2

Underactuated mechanical systems models with a tree structure actuated under mode A2 cannot be transformed into those with a chain structure. In this case, a backstepping approach cannot be employed. What can then be done to stabilize such a class of systems?

By examining the CFD of such structure Chap. 4, Fig. 4.6, it appears that the control acts at the same time on 2 DOF. Therefore, it is necessary to control the two states simultaneously. Consequently, we will include in the control law expression some terms related to the stabilization of one variable and other terms to stabilize the second one. This was the idea used to stabilize the inverted pendulum and which can be generalized to other systems of the same class.

The mathematical model of the inverted pendulum when the friction terms are neglected is given by

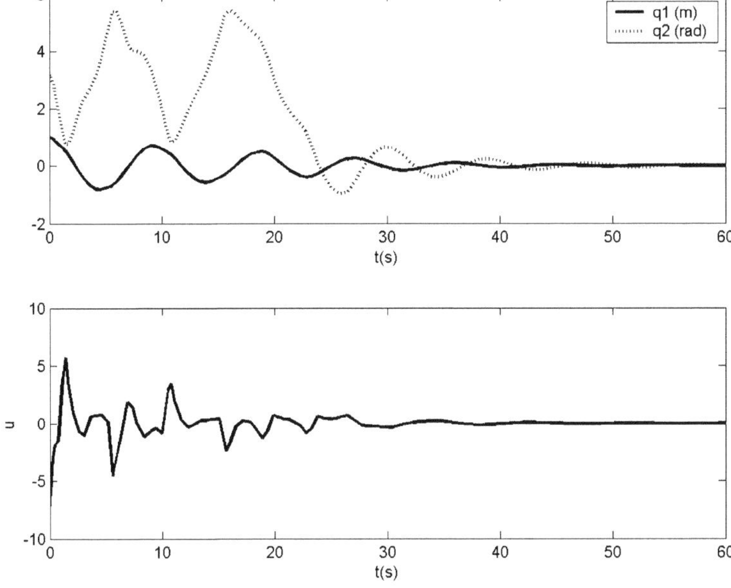

Fig. 5.13 State trajectories and control of the Tora system for the initial conditions $(q_1, q_2, \dot{q}_1, \dot{q}_2) = (1, \pi, 0, 0)$

$$ml \cos\theta \ddot{x} + \left(I + ml^2\right)\ddot{\theta} - mgl \sin\theta = 0 \tag{5.31}$$

$$(M + m)\ddot{x} + ml \cos\theta \ddot{\theta} - ml \sin\theta \dot{\theta}^2 = F \tag{5.32}$$

and from (5.31), we have

$$\ddot{x} = g \tan\theta - \frac{\left(I + ml^2\right)}{ml \cos\theta}\ddot{\theta} \tag{5.33}$$

We set $\ddot{\theta} = W_\theta = -k_1 \dot{\theta} - k_2 \theta$ for k_i, $i = 1, 2$, positive constants. Equation (5.33) becomes

$$\ddot{x} = g \tan\theta - \frac{\left(I + ml^2\right)}{ml \cos\theta}W_\theta \tag{5.34}$$

In this approach, we propose to control the pendulum through the acceleration of the cart. Thus, the expression of $\ddot{x}$ (5.34) will be considered as the necessary desired acceleration to stabilize the pendulum. Let

$$\ddot{x}_d = g \tan\theta - \frac{\left(I + ml^2\right)}{ml \cos\theta}W_\theta \tag{5.35}$$

from (5.32), we have

$$(M + m)W_x + ml \cos\theta W_\theta - ml \sin\theta \dot{\theta}^2 = F \tag{5.36}$$

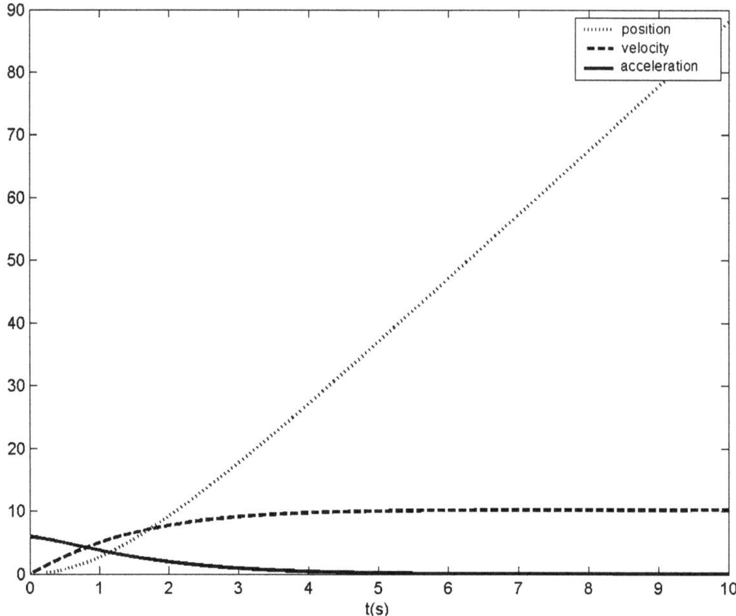

Fig. 5.14 Acceleration, velocity, and position of the cart

with

$$W_\theta = -k_1\dot\theta - k_2\theta$$
$$W_x = \ddot{x}_d - k_{x1}(\dot{x} - \dot{x}_d) - k_{x2}(x - x_d) \tag{5.37}$$

then, when θ stabilizes, $\theta \to 0$, we will have $\dot\theta \to 0$ and $\ddot\theta \to 0$. Consequently, $W_\theta \to 0$. From (5.35), we will also have $\ddot{x}_d \to 0$, but after integration, $\dot{x}$ will be a constant and x a line, which will produce the divergence of x, see Fig. 5.14.

To remedy this problem, we introduced terms to stabilize x in the expression of W_θ. In this case, (5.37) becomes

$$W_\theta = -k_1\dot\theta - k_2\theta - k_3\dot{x} - k_4 x$$
$$W_x = \ddot{x}_d - k_{x1}(\dot{x} - \dot{x}_d) - k_{x2}(x - x_d) \tag{5.38}$$

Hence, when $\theta \to 0$, $\ddot{x}_d$ is no longer zero and tends to W_θ, which in turn tends to $-k_3\dot{x} - k_4 x$ and leads x to 0.

Note that only the local stability has been proved.

The choice of the constants k_i is optimized using a pole placement procedure. For the desired spectrum $\{-1, -2, -3, -4, -5, -6\}$, the resulting gains are given by $k_1 = 14.4085$, $k_2 = 34.815$, $k_3 = 4.64$, $k_4 = 2.20$, $k_{x1} = 9.93$, $k_{x2} = 28.92$.

The application of the control law with these gains produces satisfying results as shown in Fig. 5.15. The settling time is less than 8 seconds for a large initial angle.

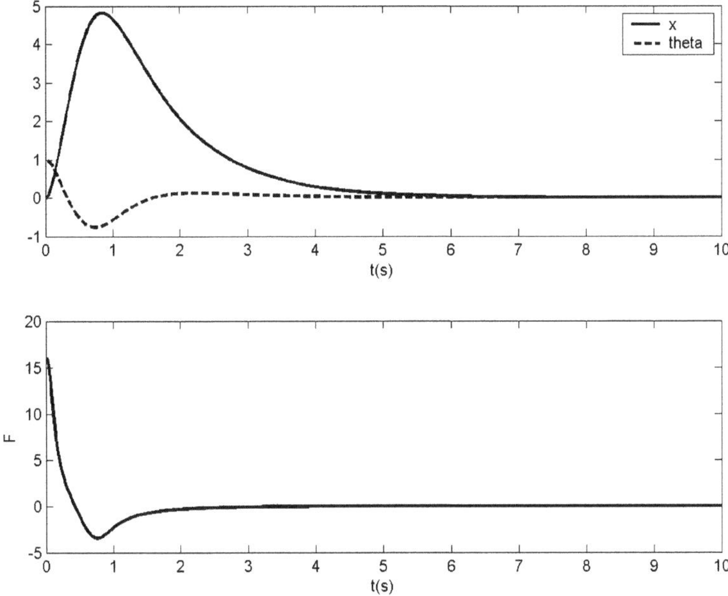

Fig. 5.15 Trajectories of the pendulum, the cart, and voltage of the motor for the initial conditions $(0, 1, 0, 0)$

The trajectories are very smooth and the control effort is acceptable. Evidently, one can still improve this control effort at the expense of the settling time by playing on the desired spectrum.

Remark 5.5 In the absence of a Lyapunov function to justify the stability of the control and the basin of attraction, one can estimate this basin by simulation. It is found to be around 1.18 radians or 67.609 degrees.

In order to make the generalization of this procedure easier, a more classical procedure can be employed. This is based on the fact that the control must contain stabilizing terms for the two variables in parallel. We use a partial linearization and then determine the expression of the control by including the stabilizing terms of the two considered degrees of freedom. This technique was applied on the inverted pendulum and it gave similar results to those of the first procedure based on the acceleration of the cart.

Let us come back to the pendulum model given by

$$ml\cos\theta\ddot{x} + \left(I + ml^2\right)\ddot{\theta} - mgl\sin\theta = 0 \tag{5.39}$$

$$(M + m)\ddot{x} + ml\cos\theta\ddot{\theta} - ml\sin\theta\dot{\theta}^2 = F \tag{5.40}$$

from (5.39), we have

$$\ddot{x} = g \tan \theta - \frac{(I + ml^2)}{ml \cos \theta} \ddot{\theta} \tag{5.41}$$

which we replace in (5.40):

$$(M + m)\left(g \tan \theta - \frac{(I + ml^2)}{ml \cos \theta} \ddot{\theta} \right) + ml \cos \theta \, W_\theta - ml \sin \theta \dot{\theta}^2 = F \tag{5.42}$$

We set

$$\ddot{\theta} = W_\theta = -k_1 \dot{\theta} - k_2 \theta - k_3 \dot{x} - k_4 x \tag{5.43}$$

The reasons of this choice were explained previously. The controlled system is given by

$$\ddot{\theta} = -k_1 \dot{\theta} - k_2 \theta - k_3 \dot{x} - k_4 x \tag{5.44}$$

$$\ddot{x} = g \tan \theta - \frac{(I + ml^2)}{ml \cos \theta}(-k_1 \dot{\theta} - k_2 \theta - k_3 \dot{x} - k_4 x) \tag{5.45}$$

The choice of the constants k_i, $i = 1, \ldots, 4$ is also done through an optimization procedure, but this time there are fewer parameters to identify. The set of parameters resulting from the desired spectrum $\{-1, -2, -3, -4\}$ is $k_1 = 15.3014$, $k_2 = 37.5447$, $k_3 = 5.0968$, $k_4 = 2.4465$.

The simulation results corresponding to these gains are given in Fig. 5.16. The results are satisfying and are practically of the same order as those obtained with a control via the cart acceleration.

Finally, we can say that UMSs actuated under the mode A2 can be stabilized through a partial linearization followed by the synthesis of a control including terms acting on the different degrees of freedom in parallel.

Now, it remains to find a solution to the problem of stabilizing the class of UMSs with an isolated vertex structure. It has been shown (Chap. 3, Sect. 4.1.1) that the systems with such structure are more difficult to control due the fact that some degrees of freedom are not influenced by the control. Essentially, this is due to the fact that the relative degree is not well-defined.

5.3 Stabilization of UMSs with an Isolated Vertex Structure

The proposed idea for the control of these systems is to use an approximate linearization approach (Appendix A) in such a way that the terms constituting an obstruction to the definition of the relative degree will be eliminated.

Nevertheless, in the case where the linearization is possible, it is perhaps not the best solution. In particular, when one modifies the system to allow this linearization or when it is subjected to parametric variations, it will lead to a non-robust system. On the other hand, a robust control such as a sliding mode control can compensate for this lack of robustness.

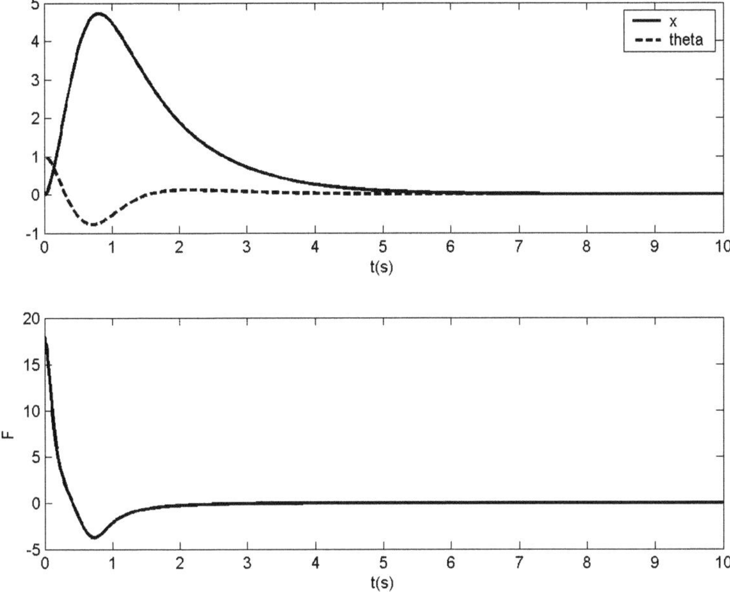

Fig. 5.16 Trajectories of the pendulum, the cart and voltage of the motor for the initial conditions $(0, 1, 0, 0)$

In order to improve this approach, we will consider a high order approximation (Appendix A) instead of a simple linear approximation. The two procedures have been applied to a system with an isolated vertex structure: the ball and beam system.

5.3.1 Control Law via Approximate Linearization

For a single input affine system:

$$\dot{x} = f(x) + g(x)u, \quad x \in \mathbb{R}^n$$
$$y = h(x) \tag{5.46}$$

The approximate linearization of this system is given by

$$\dot{z}_1 = z_2$$
$$\dot{z}_2 = z_3$$
$$\vdots \tag{5.47}$$
$$\dot{z}_{n-1} = z_n$$
$$\dot{z}_n = L_f^n h\big(\phi^{-1}(z)\big) + L_g L_f^{n-1} h\big(\phi^{-1}(z)\big)u$$

where $z = \phi(x)$ is the diffeomorphism allowing this transformation. The next lemma gives a control algorithm which forces a single input system to reach a sliding surface in finite time [20].

Lemma 5.1 [21] *Let $s(t) = s(x(t))$ be a smooth output of the single input system (5.46), with $\dot{s}(t) = a_s(x(t)) + u(t)b_s(x(t))$. Assume that the set S is defined by*

$$S = \{x \,|\, s(x) = 0\}$$

a one dimensional submanifold of $\mathbb{R}^n$ such that 0 is regular value of s. Consider a nonempty compact and convex subset D of $\mathbb{R}^n$ such that $D \cap S \neq \emptyset$. Let

$$u_*(x) = -K \operatorname{sign}(s(x)) \operatorname{sign}(b_s(x)) \tag{5.48}$$

where $K > 0$ and $K|b_s(x)| \geq |a_s(x)|$ for all $x \in D$. Then, there exists an open set $D_0 \subset D$ such that for a system starting in $x_0 \in D_0$ with $u = u_*$ it satisfies the following: there exists a $\tau > 0$ for which $x(t) \in S,\ \forall t \geq \tau$.

This result leads to the following theorem:

Theorem 5.3 [21] *Consider system (5.47), with a sliding surface:*

$$S = \{z \in \mathbb{R}^n \,|\, z_n + a_{n-1}z_{n-1} + \cdots + a_2 z_2 + a_1 z_1 = 0\} \tag{5.49}$$

such that the polynomial:

$$P(s) = s^{n-1} + a_{n-1}s^{n-2} + \cdots + a_1$$

is Hurwitz. Let D be a compact and convex neighborhood of $z = 0$ and let

$$u_{MG}^{\mathrm{app}} = -k \operatorname{sign}\big(L_g L_f^{n-1} h(\phi^{-1}(z))\big) \operatorname{sign}(s(z)) \tag{5.50}$$

with $k > 0$ and such that

$$k\big|L_g L_f^{n-1} h(\phi^{-1}(z))\big| \geq \left|\sum_{i=1}^{n-1} a_i z_{i+1} + L_f^n h(\phi^{-1}(z))\right| \quad \text{for all } z \in D,$$

then there exists an open neighborhood $D_0 \subset D$ of $z = 0$ such that system (5.47) subject to $u = u_{MG}^{\mathrm{app}}$ and $z_0 = z(0) \in D_0$ is asymptotically stable.

The proof of this theorem is given in [21].

5.3.2 Control Law via High Order Approximate Linearization

The high order approximation for the same system (5.46) is given by one of the forms

$$
\begin{aligned}
\dot{z}_1 &= z_2 + L_g h\big(\phi^{-1}(z)\big)u & \dot{z}_1 &= z_2 \\
\dot{z}_2 &= z_3 + L_g L_f h\big(\phi^{-1}(z)\big)u & \dot{z}_2 &= z_3 \\
&\;\;\vdots & &\;\;\vdots \qquad\qquad +\Theta_E^2 u \\
\dot{z}_{n-1} &= z_n + L_g L_f^{n-2} h\big(\phi^{-1}(z)\big)u & \dot{z}_{n-1} &= z_n \\
\dot{z}_n &= L_f^n h\big(\phi^{-1}(z)\big) + L_g L_f^{n-1} h\big(\phi^{-1}(z)\big)u & \dot{z}_n &= \alpha(z) + \beta(z)u
\end{aligned}
\tag{5.51}
$$

In this case the previous theorem is still valid, but one must replace u_{MG}^{app} by u_{MG} given by

$$
u_{MG} = -k\,\text{sign}\left(L_g L_f^{n-1} h\big(\phi^{-1}(z)\big) + \sum_{i=1}^{n-1} a_i L_g L_f^{i-1} h\big(\phi^{-1}(z)\big) \right) \text{sign}\big(s(z)\big)
\tag{5.52}
$$

with $k > 0$ and

$$
k\left| L_g L_f^{n-1} h\big(\phi^{-1}(z)\big) + \sum_{i=1}^{n-1} a_i L_g L_f^{i-1} h\big(\phi^{-1}(z)\big) \right| \geq \left| \sum_{i=1}^{n-1} a_i z_{i+1} + L_f^n h\big(\phi^{-1}(z)\big) \right|
$$

for all $z \in D$.

5.3.3 Application: The Ball and Beam System

Consider the beam and ball system as depicted in Fig. 5.17.

The objective of the control is to stabilize the beam on the horizontal position and the ball on the center of the beam with the application of a torque on the axis of rotation only.

The model issued from the Lagrange formalism is given by

$$
\begin{cases}
m\ddot{r} + mg\sin(\theta) - mr\dot{\theta}^2 = 0 \\
(mr^2 + I)\ddot{\theta} + 2mr\dot{r}\dot{\theta} + mgr\cos(\theta) = \tau
\end{cases}
\tag{5.53}
$$

A preliminary state feedback

$$
\tau = 2mr\dot{r}\dot{\theta} + mgr\cos(\theta) + (mr^2 + I)u
$$

Fig. 5.17 The ball and beam
system

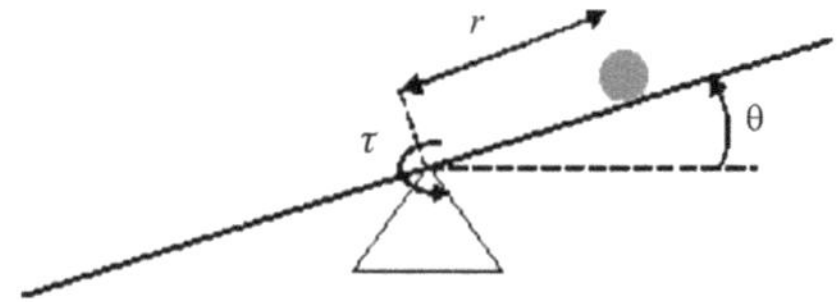

can be applied to have $\ddot{\theta} = u$. Let $(x_1, x_2, x_3, x_4) \overset{\text{def}}{=} (r, \dot{r}, \theta, \dot{\theta})$. Then, the dynamics
becomes

$$\begin{cases} \dot{x}_1 = x_2 \\ \dot{x}_2 = \mathscr{B}\left(x_1 x_4^2 - g \sin(x_3)\right) \\ \dot{x}_3 = x_4 \\ \dot{x}_4 = u \end{cases} \tag{5.54}$$

where $\mathscr{B} = 0.72$ and $g = 9.8$.

By considering the output $y = h(x) = x_1$, we have the result by direct calculation
of the Lie derivatives (Appendix C) that

$$L_g h(x) = 0$$

$$L_g L_f h(x) = 0$$

$$L_g L_f^2 h(x) = 2\mathscr{B} x_1 x_4$$

$$L_g L_f^3 h(x) = 2\mathscr{B} x_2 x_4 - \mathscr{B} g \cos(x_3)$$

Since $L_g h(x) = L_g L_f h(x) = 0$ and $= L_g L_f^2 h(x) \neq 0$, the relative degree is 3. On
the other hand, since $L_g h(0) = L_g L_f h(0) = L_g L_f^2 h(0) = 0$ and $= L_g L_f^3 h(0) \neq 0$,
the robust relative degree is 4 (Appendix A).

Using the transformation

$$\phi = \begin{cases} h(x) = x_1 = z_1 \\ L_f h(x) = x_2 = z_2 \\ L_f^2 h(x) = \mathscr{B} x_1 x_4^2 - \mathscr{B} g \sin(x_3) = z_3 \\ L_f^3 h(x) = \mathscr{B} x_2 x_4^2 - \mathscr{B} g x_4 \cos(x_3) = z_4 \end{cases}$$

and assuming $2\mathscr{B} x_1 x_4 = 0$ for an approximate linearization

$$\begin{aligned} \dot{z}_1 &= z_2 \\ \dot{z}_2 &= z_3 \\ \dot{z}_3 &= z_4 \\ \dot{z}_4 &= L_f^4 h\left(\phi^{-1}(z)\right) + L_g L_f^3 h\left(\phi^{-1}(z)\right) u \end{aligned} \tag{5.55}$$

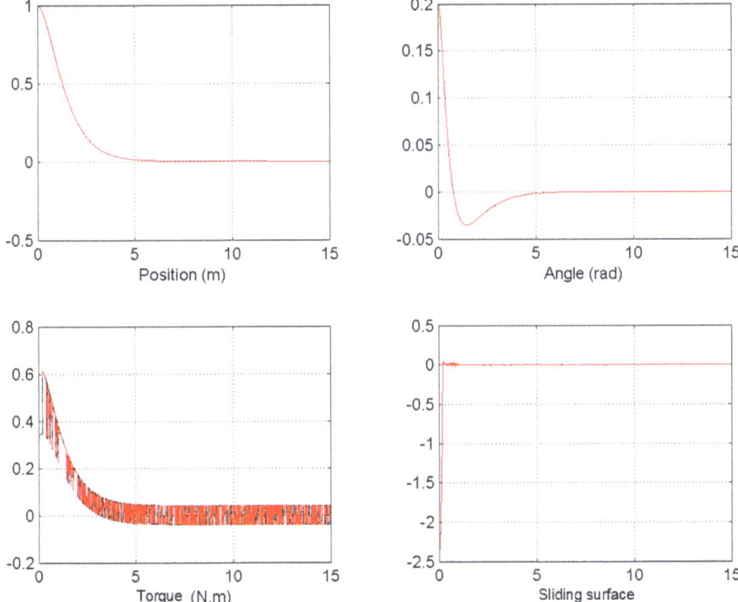

Fig. 5.18 Trajectories, torque and sliding surface for the ball and beam system for the initial conditions: $x_{01} = 1$ m, $x_{03} = 0.2$ rad

while assuming $2\mathscr{B}x_1x_4 \neq 0$ for a high order approximate linearization:

$$
\begin{aligned}
\dot{z}_1 &= z_2 \\
\dot{z}_2 &= z_3 \\
\dot{z}_3 &= z_4 + L_g L_f^2 h\big(\phi^{-1}(z)\big) \\
\dot{z}_4 &= L_f^4 h\big(\phi^{-1}(z)\big) + L_g L_f^3 h\big(\phi^{-1}(z)\big)u
\end{aligned}
\tag{5.56}
$$

The expressions of the control law for (5.55) and (5.56) for the same sliding surface: $S = \{z \in \mathbb{R}^4 \mid z_4 + a_3 z_3 + a_2 z_2 + a_1 z_1 = 0\}$ where $a_3 = 6$, $a_2 = 12$, and $a_1 = 8$ are, respectively, given by

$$
u_{MG}^{\text{app}} = -k \times \text{sign}\big(L_g L_f^3 h\big(\phi^{-1}(z)\big)\big) \times \text{sign}\big(s(z)\big)
\tag{5.57}
$$

$$
u_{MG} = -k \times \text{sign}(L_g L_f^3 h\big(\phi^{-1}(z)\big) + L_g L_f^2 h\big(\phi^{-1}(z)\big)\,\text{sign}\big(s(z)\big)
\tag{5.58}
$$

Simulation results show that the two control laws, when applied to the ball and beam, are of equal efficiency for small initial conditions, see Fig. 5.18, whereby all the curves are superposed to each other and tend to 0.

As soon as the initial conditions become more important to enlarge the attraction domain, one can remark from Fig. 5.19 that the first control, u_{MG}^{app}, is no longer sufficient to stabilize the system while the second one, u_{MG}, remains valid.

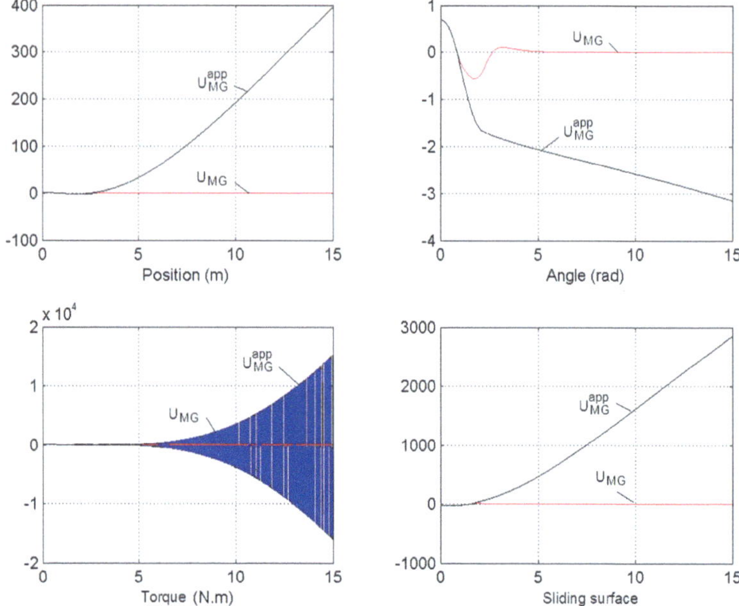

Fig. 5.19 Trajectories, torque, and sliding surface for the ball and beam system for the initial conditions: $x_{01} = 1$ m, $x_{03} = 0.7$ rad related to the controls u_{MG}^{app} and u_{MG}

Consequently, the higher order approximation permits to enlarge the stability domain.

Nevertheless, the two control laws can generate chattering due to the discontinuous nature of the control by sliding mode, which can damage the motor.

To remedy this problem, two solutions can be proposed: The first one consists in using a more regular function than the sign function, for example the arctangent function [5]. This modification cannot only reduce the chattering Fig. 5.20 but can also permit to enlarge the basin of attraction with respect to a sign function and this even when a high order approximation is not employed, see Fig. 5.21.

The second solution consists in synthesizing a second order sliding mode control law such as the twisting or the super twisting algorithms.

5.4 Summary

In this chapter, the stabilization problems of most of the UMSs with two degrees of freedom have been considered. The analysis of the CFD of a system constitutes the starting point of the procedure. Depending on the obtained CFD, different approaches leading to simple control law design that are easy to implement are proposed. To summarize, one can say that UMSs possessing a CFD in chain form can be stabilized with a systematic backstepping procedure. For UMSs possessing a CFD

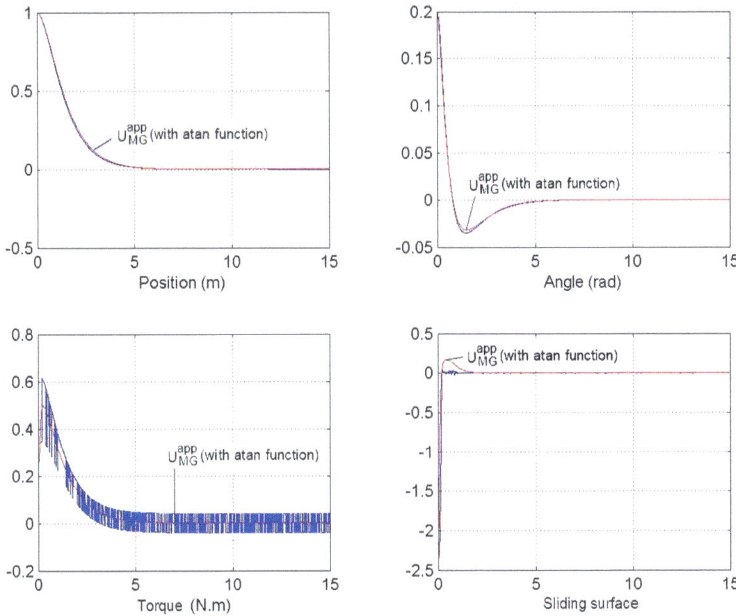

Fig. 5.20 Trajectories, torque, and sliding surface related to the ball and beam system for the initial conditions: $x_{01} = 1$ m, $x_{03} = 0.2$ rad when the function sign is replaced by an Atan function

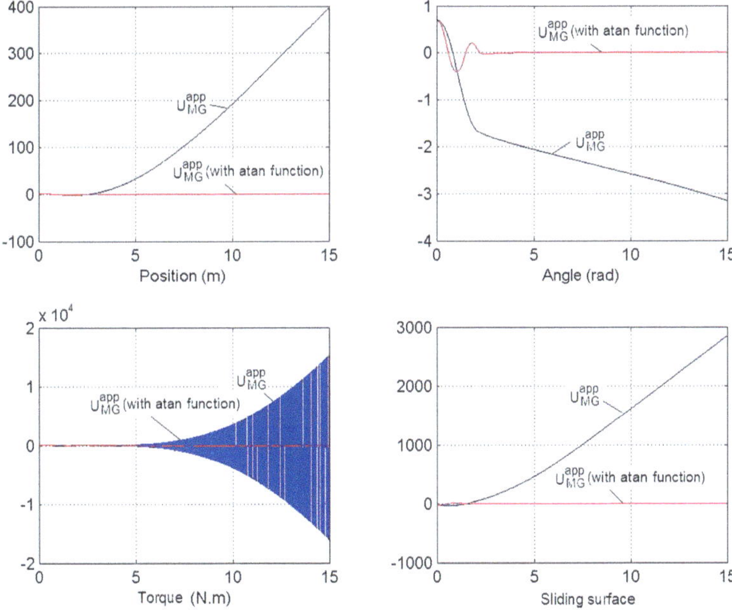

Fig. 5.21 Trajectories, torque, and sliding surface related to the ball and beam system for the initial conditions: $x_{01} = 1$ m, $x_{03} = 0.7$ rad for the control law u_{MG}^{app} and for the control law u_{MG}^{app} in which the sign function is replaced by an Atan function

with tree structure, the idea is to verify the actuation mode of the variable appearing in the inertia matrix. Thus, when this variable, called shape variable, is actuated, the CFD in a tree structure can be transformed into a CFD with chain structure allowing the use of the systematic backstepping procedure. However, due to this transformation, very often some parts of the hypotheses that ensure the global stability are no longer verified, due to the fact that the control law possesses singularities at some points. In this case, two solutions to bypass the singularities are proposed. The first one consists in employing a switching control depending on the domain of the singularity. The second solution consists in completely changing the control law and acting on the gains so as to force one condition to hold true. These two solutions allow a global stability despite the singularities in the control. Consequently, one of the hypotheses of the *Seto and Baillieul* control scheme is relaxed. Next, when the shape variable is not actuated, this implies that the associated systems cannot be put in the form of a chain structure. In this case, a procedure based on partial linearization by including stabilizing terms of the two degrees of freedom in the control law in parallel was proposed. For the UMSs possessing a CFD with an isolated vertex structure, the control objectives can be achieved through an approximate linearization or eventually by high order approximations followed by a robust control of sliding mode type. All these control procedures have been tested on examples of UMSs, and the simulation results confirmed their efficiencies.

References

1. P. Baranyi, P.L. Varkonyi, Exact tensor product distributed compensation based stabilization of the Tora system, in *WSEAS Int. on Dynamical Systems and Control*, Italy (2005), pp. 38–43
2. R.W. Brockett, *Asymptotic Stability and Feedback Stabilization* (Birkhäuser, Basel, 1983)
3. A. Choukchou-Braham, B. Cherki, A new control scheme for a class of underactuated systems. Mediterr. J. Meas. Control **7**(1), 204–210 (2011)
4. A. Choukchou-Braham, B. Cherki, M. Djemai, Stabilisation of a class of underactuated system with tree structure by backstepping approach. Nonlinear Dyn. Syst. Theory **12**(4), 345–364 (2012)
5. A. Choukchou-Braham, C. Bensalah, B. Cherki, Stabilization of an under-actuated mechanical system by sliding control, in *Proc. 1st Conf. on Intelligent Systems and Automation CISA'09* (American Institute of Physics AIP, New York, 2008), pp. 80–84
6. A.D. Luca, Dynamic control of robots with joint elasticity, in *Proc. 31st IEEE Conf. on Robotics and Automation* (1988), pp. 152–158
7. J.M. Flaus, *Stabilité des Systèmes Dynamiques Hybrides* (Hermes, London, 2001)
8. A.L. Frafkov, I.A. Makarov, A. Shiriaev, O.P. Tomchina, Control of oscillations in Hamiltonian systems, in *Proc. European Control Conf.*, Brussels (1997)
9. G. Hancke, A. Szeghegyi, Nonlinear control via TP model transformation: the Tora system example, in *Symposium on Applied Machine Intelligence*, Slovakia (2004)
10. J. Huang, G. Hu, Control design for the nonlinear benchmark problem via the output regulation method. J. Control Theory Appl. **2**, 11–19 (2004)
11. L. Hung, H. Lin, H. Chung, Design of self tuning fuzzy sliding mode control for Tora system. Expert Syst. Appl. **32**(1), 201–212 (2007)
12. M. Jankovic, D. Fontaine, P.V. Kokotović, Tora example: cascade and passivity-based control designs. IEEE Trans. Control Syst. Technol. **4**(3), 292–297 (1996)

13. Z.P. Jiang, I. Kanellakapoulos, Global output feedback tracking for a benchmark nonlinear system, in *Proc. 31st IEEE Conf. on Robotics and Automation*, USA (1999), pp. 4802–4807

14. S. Nazrulla, H.K. Khalil, A novel nonlinear output feedback control applied to the Tora benchmark system, in *Proc. 47th IEEE Conf. on Decision and Control*, Mexico (2008), pp. 3565–3570

15. R. Olfati Saber, Nonlinear control of underactuated mechanical systems with application to robotics and aerospace vehicles. Ph.D. thesis, Massachusetts Institute of Technology, Department Electrical Engineering and Computer Science, 2001

16. N. Qaiser, A. Hussain, N. Iqbal, N. Qaiser, Dynamic surface control for stabilization of the oscillating eccentric rotor, in *IMechE Part Systems and Control Engineering* (2007), pp. 311–319

17. R. Sepulchre, M. Janković, P. Kokotović, *Constructive Nonlinear Control* (Springer, Berlin, 1997)

18. D. Seto, J. Baillieul, Control problems in super articulated mechanical systems. IEEE Trans. Autom. Control **39**(12), 2442–2453 (1994)

19. D. Seto, A.M. Annaswamy, J. Baillieul, Adaptive control of a class of nonlinear systems with a triangular structure. IEEE Trans. Autom. Control **39**(7), 1411–1428 (1994)

20. S.K. Spurgeon, C. Edwards, *Sliding Mode Control: Theory and Applications* (Taylor and Francis, London, 1983)

21. D.A. Voytsekhovsky, R.M. Hirschorn, Stabilization of single-input nonlinear systems using higher order compensating sliding mode control, in *Proc. 45th IEEE Conf. on Decision and Control and the European Control Conf.* (2005), pp. 566–571

22. C.J. Wan, D.S. Bernstein, V.T. Coppola, Global stabilization of the oscillating eccentric rotor. Nonlinear Dyn. **10**, 49–62 (1996)

23. H. Wang, J. Li, *Nonlinear Control Via PDC: The Tora System Example* (2003)

24. R. Xu, U. Ozguner, Sliding mode control of a class of underactuated systems. Automatica **44**, 233–241 (2008)

Appendix A
Theoretical Background of Nonlinear System Stability and Control

> *... Mr. Fourier had the opinion that the main purpose of*
> *mathematics was public utility and explanation of natural*
> *phenomena, but a philosopher like him should have known that*
> *the sole purpose of science is the honor of the human spirit, and*
> *that, as such, a matter of numbers is worth as much as a matter*
> *of the world system.*
>
> *Letter from Jacobi Legendre, July 2nd, 1830*

Automatic control comprises a number of theoretical tools of mathematical characteristics that enable to predict and apply its concepts to fulfill the objectives that are directly attached to it. These tools are necessary for the synthesis of control laws on a specific process and are utilized at various stages of the design. This is more so, particularly during the modeling and identification of the parameters stage as well as during the construction of control laws and during the verification of the stability of the controlled system, just to mention a few. In fact, it is well-known that all construction techniques of control laws or observation are narrowly linked to stability considerations.

As a result, in the first part of this appendix, some definitions and basic concepts of stability theory are recalled. The second part of the appendix is dedicated to the presentation of some concepts and techniques of control theory.

Due to the numerous contributions in this area, in the past few years, we have focused our interest only on points that are more directly related to our own work.

A.1 Stability of Systems

One of the tasks of the control engineer consists, very often, in the study of stability, whether for the considered system, free from any control, or for the same system when it is augmented with a particular control structure. At this stage, it might be useful or even essential to ask what stability is. How do we define it? How to conceptualize and formalize it? What are the criteria upon which one can conclude on the stability of a system?

A. Choukchou-Braham et al., *Analysis and Control of Underactuated Mechanical Systems*, DOI 10.1007/978-3-319-02636-7,
© Springer International Publishing Switzerland 2014

A.1.1 What to Choose?

It is clear that drawing up an inventory as complete as possible of the forms of stability that have appeared throughout the history of automatic control but also of mechanics would be beyond the scope of this book. There will therefore not be included in this presentation the stability method of Krasovskii [70], comparison method, singular perturbations [37], the stability of the UUB (Uniformly Ultimately Bounded) [15], the input–output stability [94], the input to state stability [71], the stability of non-autonomous systems [2], contraction analysis [34], and descriptive functions [70].

In addition, we will not be presenting the proofs of various results in this section. We will assume that the conditions of existence and the uniqueness of solutions for the considered systems of differential equations are verified everywhere.

From a notation point of view, we shall denote by $x(t, t_0, x_0)$ the solution at time t with initial condition x_0 at time t_0 or by $x(t, t_0, x_0, u)$ when the system is controlled. In addition, for simplicity, we shall frequently use the notation $x(t)$ or even x, when the dependence on t_0, x_0 or t is evident. Similarly, we shall consider in the majority of cases, except in some cases, the initial time $t_0 = 0$.

The class of systems considered will be those that can be put in the following ordinary differential equation (ODE) form:

$$\dot{x} = f(x) \tag{A.1}$$

where $x \in \mathbb{R}^n$ is the state vector and $f : D \to \mathbb{R}^n$ a locally Lipschitzian function and continuous on the subset D of $\mathbb{R}^n$.

This type of systems is also called autonomous due to the absence of the temporal term t in the function. Non-autonomous systems of the form $\dot{x} = f(x, t)$ are not considered in this work.

For the above equation (A.1), the point of the state space $x = 0$ is an equilibrium point if it verifies

$$f(0) = 0 \quad \forall t \geq 0 \tag{A.2}$$

Note that, by a change of coordinates, one can always bring the equilibrium point to the origin.

A.1.2 The Lyapunov Stability Theory

The Lyapunov stability theory is considered as one of the cornerstones of automatic control and stability for ordinary differential equations in general. The original theory of Lyapunov dates back to 1892 and deals with the study of the behavior of solution of differential equation for different initial conditions. One of its applica-

Fig. A.1 Intuitive illustration of stability

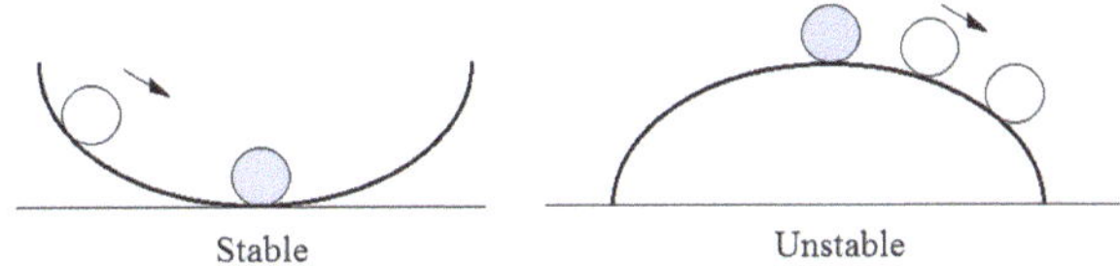

tions that was contemplated at that time was the study of librations in astronomy.[1][2] The emphasis is focused on the ordinary stability (i.e. stable but not asymptotically stable), which we can represent as a robustness with respect to initial conditions, and the asymptotic stability is only addressed in a corollary manner.

The automatic control community having inverted this preference, we will be concentrating here on the concept of asymptotic stability rather than mere stability.

Note that there are many more complete presentations of Lyapunov stability in many articles, for example [37, 49, 55, 62, 65, 66, 70, 87], which constitute the main references of this part; this list also does not claim to be exhaustive.

A.1.2.1 Stability of Equilibrium Points

Roughly speaking, we say that a system is stable if when displaced slightly from its equilibrium position, it tends to come back to its original position. On the other hand, it is unstable if it tends to move away from its equilibrium position (see Fig. A.1).

Mathematically speaking, this is translated into the following definitions:

Definition A.1 [37]   The equilibrium point $x = 0$ is said to be:

- *stable*, if for every $\varepsilon > 0$, there exists $\eta > 0$ such that for every solution $x(t)$ of (A.1) we have

$$\|x(0)\| < \eta \quad \Rightarrow \quad \|x(t)\| < \varepsilon \quad \forall t \geq 0$$

- *unstable*, if it is not stable, that is, if for every $\varepsilon > 0$, there exists $\eta > 0$ such that for every solution $x(t)$ of (A.1) we have

$$\|x(0)\| < \eta \quad \Rightarrow \quad \|x(t)\| \geq \varepsilon \quad \forall t \geq 0$$

- *attractive*, if there exists $r > 0$ such that for every solution $x(t)$ of (A.1) we have

$$\|x(0)\| < r \quad \Rightarrow \quad \lim_{t \to \infty} x(t) = 0$$

The basin of attraction of the origin is defined by the set $\mathbb{B}$ such that

$$x(0) \in \mathbb{B} \quad \Rightarrow \quad \lim_{t \to \infty} x(t) = 0$$

---

[1] In astronomy, the librations are small oscillations of celestial bodies around their orbits.

[2] The father of Alexander Michael Lyapunov was an astronomer.

Fig. A.2 Stability and
asymptotic stability of $\bar{x}$

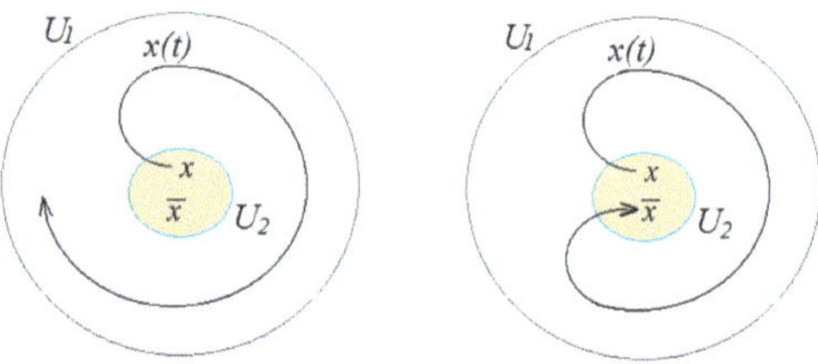

- *globally attractive*, if for every solution $x(t)$ of (A.1) we have

$$\lim_{t \to \infty} x(t) = 0. \quad \text{In this case,} \quad \mathbb{B} = \mathbb{R}^n$$

- *asymptotically stable*, if it is stable and attractive, and *globally asymptotically stable* (*GAS*), if it is stable and globally attractive.
- *exponentially stable*, if there exist $r > 0$, $M > 0$ and $\alpha > 0$ such that for every solution $x(t)$ of (A.1) we have

$$\left\| x(0) \right\| < r \quad \Rightarrow \quad \left\| x(t) \right\| \leq M \left\| x(0) \right\| e^{-\alpha t} \quad \text{for all } t \geq 0$$

and *globally exponentially stable* (*GES*), if there exist $M > 0$ and $\alpha > 0$ such that for every solution $x(t)$ of (A.1) we have

$$\left\| x(t) \right\| \leq M \left\| x(0) \right\| e^{-\alpha t} \quad \text{for all } t \geq 0$$

Remark A.1

1. The difference between stable and asymptotically stable is that a small perturbation on the initial state around a stable equilibrium point $\bar{x}$ might lead to small sustained oscillations, whereas these oscillations are dampened in time in the case of asymptotically stable equilibrium point, see Fig. A.2 (U_1 is the ball of center 0 and radius ε and U_2 is the ball of center 0 and radius η [46]).
2. For a linear system, all these definitions are equivalent (except for stable and asymptotically stable). However, for a nonlinear system, stable does not imply attractive, attractive does not imply stable, asymptotically stable does not imply exponentially stable whereas exponentially stable implies asymptotically stable.

When the systems are represented by nonlinear differential equations, the verification of stability is not trivial. On the contrary, for linear systems the verification of stability is systematic and is determined as follows.

A.1.2.2 Stability of the Origin for a Linear System

Consider the linear system

$$\dot{x} = Ax \tag{A.3}$$

where A is a square matrix of dimension n. Let $\lambda_1, \ldots, \lambda_s$, be the distinct eigenvalues with algebraic multiplicity $m(\lambda_i)$ of the matrix A.

Theorem A.1 [37]

1. *If* $\exists j\ \mathrm{Re}(\lambda_j) > 0$ *or if* $\exists k\ \mathrm{Re}(\lambda_k) = 0$ *and* $m(\lambda_k) > 1$ *then* $x = 0$ *is unstable.*
2. *If* $\forall j\ \mathrm{Re}(\lambda_j) < 0$ *then* $x = 0$ *is exponentially* (*hence asymptotically*) *stable.*
3. *If* $\mathrm{Re}(\lambda_j) < 0$ *and if* $\exists k\ \mathrm{Re}(\lambda_k) = 0$ *and* $m(\lambda_k) = 1$ *then* $x = 0$ *is stable but not attractive.*

Unfortunately, there does not exist an equivalent theorem to that of eigenvalues for nonlinear systems. In some cases, one can characterize the stability of the origin via the study of the linearized system.

A.1.2.3 Linear Approximation of a System

Consider a system of the form (A.1); we denote by

$$A = \frac{\partial f}{\partial x}(\bar{x})$$

the Jacobian matrix of f evaluated at the equilibrium point $x = \bar{x}$. The obtained system will be of the form (A.3) and is called the linearization (or linear approximation) of the nonlinear system (A.1).

Theorem A.2 [37]

1. *If* $x = 0$ *is asymptotically stable for* (A.3) *then* $x = \bar{x}$ *is asymptotically stable for* (A.1).
2. *If* $x = 0$ *is unstable for* (A.3) *then* $x = \bar{x}$ *is unstable for* (A.1).
3. *If* $x = 0$ *is stable but not asymptotically stable for* (A.3) *then we cannot conclude on the stability of* $x = \bar{x}$ *for* (A.1).

Another criterion that allows to conclude on the stable behavior of the system for both linear and nonlinear system is described next.

A.1.2.4 Lyapunov's Direct Method

The principle of this method is a mathematical extension of the following physical phenomenon: if the total energy (of positive sign) of a mechanical or electrical system is continuously decreasing then the system tends to reach a minimal energy configuration. In other words, to conclude on the stability of a system, it suffices to examine the variations of a certain scalar function called Lyapunov function without having to solve the explicit solution of the system. This is precisely the strong point of this method since the equation of motion of $x(t)$ does not have to be computed in order to characterize the evolution of the solution (the determination of explicit solutions of nonlinear system is difficult and sometimes impossible).

Lyapunov Function   Consider the system

$$\dot{x} = f(x) \quad \text{with } f(0) = 0 \tag{A.4}$$

$x = 0$ is an equilibrium point for (A.4) and $D \subset \mathbb{R}^n$ is a domain that contains $x = 0$.

Let $V : D \to \mathbb{R}$ be a function that admits continuous partial derivatives. We denote

$$\dot{V}(x) = \frac{\partial V(x)}{\partial x} \cdot f(x) = \sum_{i=1}^{n} \frac{\partial V(x)}{\partial x_i} \cdot f_i(x)$$

the derivative of the function V in the direction of the vector field f.

Definition A.2   The function V is a Lyapunov function for system (A.4) at $x = 0$ in D, if for all $x \in D$ we have

- $V(x) > 0$ except at $x = 0$ where $V(0) = 0$
- $\dot{V}(x) \leq 0$.

Theorem A.3 [37]

1. *If there exists a Lyapunov function for (A.4) at $x = 0$ in a neighborhood D of 0, then $x = 0$ is stable.*
2. *If, in addition, $x \neq 0 \Rightarrow \dot{V}(x) < 0$ then $x = 0$ is asymptotically stable.*
3. *If, in addition, $D = \mathbb{R}^n$ and $V(x) \to \infty$ when $\|x\| \to \infty$ then $x = 0$ is GAS.*

Remark A.2

1. $\dot{V}$ depends only on x, it is sometimes called the derivative of V along the system.
2. This derivative is also called Lie derivative and is denoted by $L_f V$.
3. To calculate $\dot{V}$, we do not require the knowledge of x but of $\dot{x}$, that is, of $f(x)$. Hence, for the same function $V(x)$, $\dot{V}$ is different for different systems.
4. For every solution $x(t)$ of (A.4), we have $\frac{d}{dt} V(x(t)) = \dot{V}(x(t))$, consequently if $\dot{V}$ is negative, V decreases along the solution of (A.4) so that the trajectories converge towards the minimum of V.
5. When $V(x) \to \infty$ whenever $\|x\| \to \infty$, $V(x)$ is said to be radially unbounded.
6. $V(x)$ is often a function that represents the energy or a certain form of energy of the system.
7. From a geometric point of view, a Lyapunov function is seen as a bowl whose minimum coincides with the equilibrium point. If this point is stable, then the velocity vector $\dot{x}$ (or f), tangent to every trajectory will point towards the interior of the bowl, see Fig. A.3 [46].

LaSalle's Invariance Principle

Definition A.3   A set $G \subseteq \mathbb{R}^n$ is said to be positively invariant if every solution $x(t)$ such that $x(0) \in G$ remains in G for all $t \geq 0$.

Fig. A.3 Lyapunov
function V for vector fields f

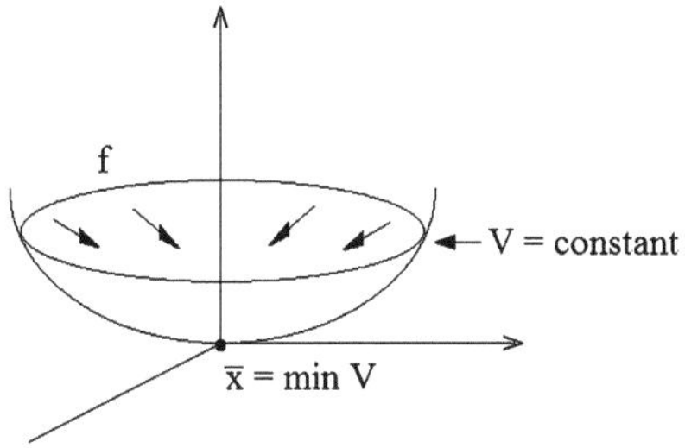

If $\bar{x}$ is an equilibrium point then $\{\bar{x}\}$ is positively invariant.

Theorem A.4 ([37] (Lyapunov–LaSalle)) *Let $V : D \to \mathbb{R}^+$ be a function having continuous partial derivatives such that there exists l for which the region D_l defined by $V(x) < l$ is bounded $\dot{V}(x) \leq 0$ for all $x \in D_l$. Let $R = \{x \in D_l : \dot{V}(x) = 0\}$ and let M be the largest positively invariant set that is included in R. Then, every solution issued from D_l tends to M when $t \to \infty$. In particular if $\{0\}$ is the only orbit contained in R then $x = 0$ is asymptotically stable and D_l is contained in its basin of attraction.*

Theorem A.5 [37] *Let $V : \mathbb{R}^n \to \mathbb{R}^+$ be a function having continuous partial derivatives. Suppose that $V(x)$ is radially unbounded and that $\dot{V}(x) \leq 0$ for all $x \in \mathbb{R}^n$. Let $R = \{x \in \mathbb{R}^n : \dot{V}(x) = 0\}$ and let M be the largest positively invariant set that is included in R. Then, every solution tends to M when $t \to \infty$. In particular if $\{0\}$ is the only orbit contained in R then $x = 0$ is GAS.*

Remark A.3

1. The criteria for stability and asymptotic stability presented in Theorems A.3, A.4 and A.5 are easy to utilize. However, they do not give any information on how to construct the Lyapunov function. In reality, there does not exist any general method for the construction of Lyapunov functions except for some particular classes of systems (namely for the class of linear systems).
2. The theorems given previously give sufficient conditions in the sense that if for a certain Lyapunov function V, the conditions on $\dot{V}$ are not satisfied, this does not imply that the considered system is unstable (maybe with another function one can demonstrate the stability of the system).
3. Contrary to Lyapunov functions which guarantee the stability of the equilibrium points, there are functions, called Chetaev functions, that guarantee the instability of the equilibrium points. Note that it is more difficult to demonstrate the instability rather than stability (refer to [46] for more details).

In some cases, a dynamical system is represented, at a given instant of time $t \geq t_0$, not by a single set of continuous differential equations, but by a family of continuous subsystems together with a logic orchestrating the switching between these subsystems: this is the class of switching systems.

In this book we have employed some controllers for this class of systems. Consequently, in what follows, we shall present the stability criteria for these systems. We shall present the controller design for this class of system for a later stage.

A.1.3 Stability of Switching Systems

Mathematically speaking, a switching system can be described by equations of the form

$$\dot{x} = f_p(x) \qquad (A.5)$$

where $\{f_p : p \in \mathbb{P}\}$ is a family of functions sufficiently regular defined from $\mathbb{R}^n$ to $\mathbb{R}^n$ and parameterized by a set of indices $\mathbb{P}$.

For the system (A.5), the active subsystem at every instant of time is determined by a sequence of switches of the form

$$\sigma = \big((t_0, p_0), (t_1, p_1), \ldots, (t_k, p_k), \ldots\big) \quad (t_0 \leq t_1 \leq \cdots \leq t_k)$$

σ is called the switching signal and can depend either on time or on the state or both. Such systems are said to have variable structures or are called multi-models. They represent a particularly simple class of hybrid systems [10, 81, 85].

Here, we shall assume that the origin is an equilibrium point that is common for the individual subsystems $f_p(0) = 0$. We shall also assume that the switching is done without jumps and does not occur infinitely fast so that the Zeno phenomenon is avoided. The reader who is interested in these properties can refer to [6, 53, 63, 64].

The class of systems often considered in the literature are those for which the individual systems are linear

$$\dot{x} = A_p x \qquad (A.6)$$

Just to mention a few, we cite the following references: [11, 26, 51, 50, 52, 54, 68, 75, 74, 90, 95, 96, 97]. On the other hand, there are only few works in the literature for the class of nonlinear switching systems [9, 12, 16, 19, 47, 53, 98, 99].

At this stage one might ask the following question: given a switching system, why do we need a theory of stability that is different from that of Lyapunov?

The main reason is that the stability of switching systems depends not only on the different dynamics corresponding to several subsystems but also on the transition law that governs the switchings. In effect, we have the case where two subsystems are exponentially stable while the switching between the two subsystems drives the trajectories to infinity.

In fact, it was shown in [12, 19, 47] that a necessary condition for the stability of switching systems subjected to an arbitrary transition law is that all the individual subsystems should be asymptotically stable, but this condition was not sufficient. Nevertheless, it appears that when the switching between the subsystems is sufficiently slow (so as to allow the transition period to settle down and to allow each subsystems to be in steady state) then it is very likely that the global system will be stable.

A.1.3.1 Common Lyapunov Function

It is clear that in the case where the family of systems (A.5) possesses a common Lyapunov function $V(x)$ such that $\nabla V(x) f_p(x) < 0$ for all $x \neq 0$ and all $p \in \mathbb{P}$, then the switching system is asymptotically stable for any transition signal σ [47]. Hence, a possibility for demonstrating the stability of switching systems consists in finding a common Lyapunov function for all the individual subsystems of (A.5).

However, finding a Lyapunov function for a nonlinear system, even for a single one is not simple. If, in addition, we allow the switchings between several subsystems, the determination of such a function becomes much more difficult. It is also the reason for which a non-classical theory of stability is necessary.

A.1.3.2 Multiple Lyapunov Functions

In the case where a common Lyapunov function cannot be determined, the idea is to demonstrate the stability through several Lyapunov functions. One of the first results of such procedure was developed by Peleties in [58, 59], then by Liberzon [47], for the switching systems of the form (A.6).

Given N dynamical systems $\Sigma_1, \ldots, \Sigma_N$, and N pseudo Lyapunov functions (Lyapunov-like functions) $V_1, \ldots, V_N$.

Definition A.4 [19]   A pseudo Lyapunov function for system (A.5) is a function $V_i(x)$ with continuous partial derivatives defined on a domain $\Omega_i \subset \mathbb{R}^n$, satisfying the following conditions:

- V_i is positive definite: $V_i(x) > 0$ and $V_i(0) = 0$ for all $x \neq 0$.
- $\dot{V}$ is semi negative definite: for $x \in \Omega_i$,

$$\dot{V}_i(x) = \frac{\partial V_i(x)}{\partial x} f_i(x) \leq 0 \tag{A.7}$$

and Ω_i is the set for which (A.7) holds true.

Theorem A.6 [19]   *Suppose that $\bigcup_i \Omega_i = \mathbb{R}^n$. For $i < j$, let $t_i < t_j$ be the transition instants for which $\sigma(t_i) = \sigma(t_j)$ and suppose that there exists $\gamma > 0$ such that*

$$V_{\sigma(t_j)}\big(x(t_{j+1})\big) - V_{\sigma(t_i)}\big(x(t_{i+1})\big) \leq -\gamma \big\| x(t_{i+1}) \big\|^2. \tag{A.8}$$

Then, the system (A.6) with $f_{\sigma(t)}(x) = A_{\sigma(t)} x$ and the transition function $\sigma(t)$ is GAS.

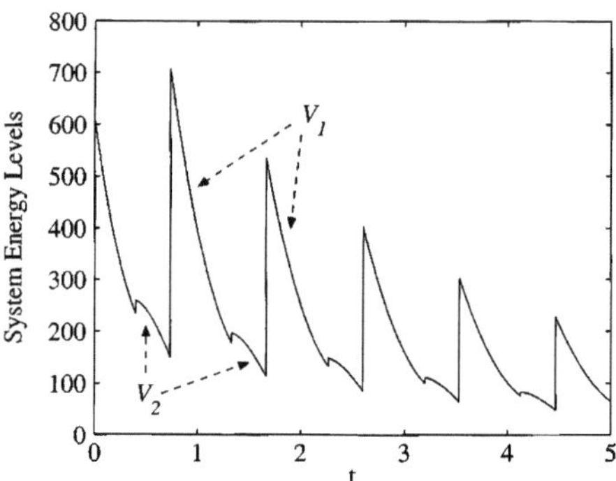

Fig. A.4 Energy profile of the linear switching system for $N = 2$

The condition (A.8) is illustrated by Fig. A.4.

The first generalization of this theorem to nonlinear systems is due to Branicky [9, 10, 11, 12]

Theorem A.7 [9, 10]   *Given N switching systems of the form* (A.5) *and N pseudo Lyapunov functions V_i in the region Ω_i associated to each subsystem, and suppose that $\bigcup_i \Omega_i = \mathbb{R}^n$ and let $\sigma(t)$ be the transition sequence that takes the value i when $x(t) \in \Omega_i$. If in addition,*

$$V_i\big(x(t_{i,k})\big) \leq V_i\big(x(t_{i,k-1})\big) \tag{A.9}$$

where $t_{i,k}$ is the kth time where f_i is active, that is, $\sigma(t_{i,k}^-) \neq \sigma(t_{i,k}^+) = i$, then (A.5) *is stable in the sense of Lyapunov.*

Figure A.5 illustrates the condition (A.9) (in dotted lines) [19]. A more general result due to Ye [91, 92] concerns the utilization of weak Lyapunov functions for which condition (A.9) is replaced by

$$V_i\big(x(t)\big) \leq h\big(V_i\big(x(t_j)\big)\big), \quad t \in (t_j, t_{j+1}) \tag{A.10}$$

where $h : \mathbb{R}^+ \to \mathbb{R}^+$ is a continuous function with $h(0) = 0$ and t_j is any transition instant when the system i is activated.

In this case, it is no longer required that the Lyapunov functions be decreasing. It suffices that they are bounded by a function that is zero at the origin. Hence, the energy can grow in the intervals where the same system is activated but must be decreasing at the end of these intervals, see Fig. A.5 (solid lines).

Liberzon in [47] extends these results by giving a condition on multiple Lyapunov functions in order to demonstrate the global asymptotic stability.

Consider N subsystems of the form (A.5). When the subsystems of the family (A.5) are assumed to be asymptotically stable, then there exists a family of Lyapunov functions $\{V_p : p \in \mathbb{P}\}$ such that the value of V_p decreases on each interval for which the pth subsystem is active.

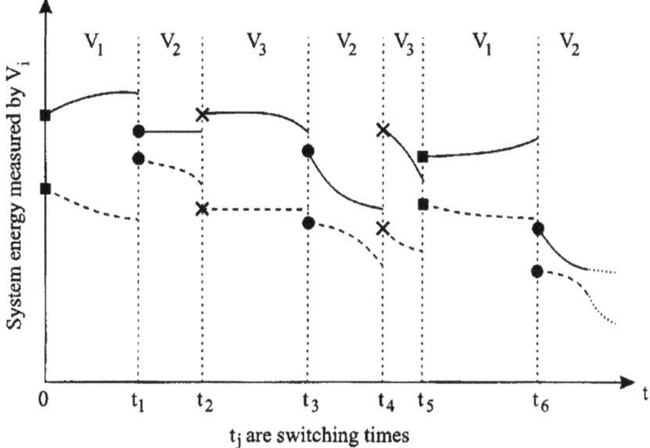

Fig. A.5 Energy profile of a nonlinear switching system for $N = 3$

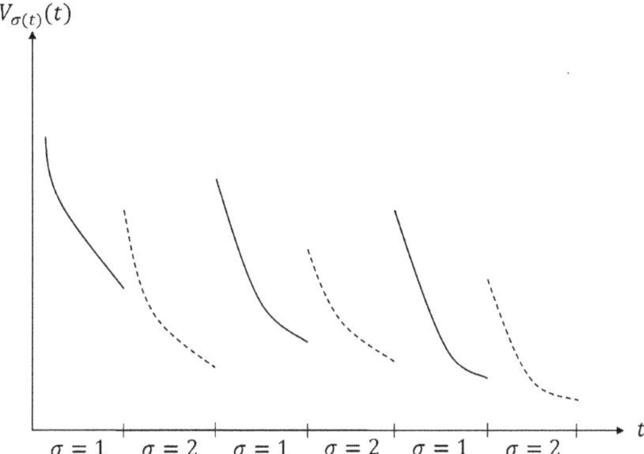

Fig. A.6 Energy profile of the nonlinear switching system for $N = 2$

If for each p, the value of V_p at the end of the interval where the system p is active is higher than the value of V_p at the end of the following interval when the system p is active (see Fig. A.6), then the system (A.5) is asymptotically stable.

Remark A.4

1. When $N = 1$, we obtain the classical results of stability. However, when $N = \infty$ the previous theorems are no longer valid.
2. These theorem are valid even when f_p vary as a function of time.
3. These results can be extended by relaxing certain hypotheses, for example: the individual subsystems can have different equilibrium points [53] or state jumps during a switch [64].

Note that all the results of stability using multiple Lyapunov functions are concerned with the decrease of these functions either at the beginning or at the end of successive intervals where the same subsystem is active. Zhai in [98] has shown that certain Lyapunov functions may not decrease at the beginning or at the end of these intervals and yet decrease globally. His demonstration, which establishes a complementary condition of stability to those that already exist, is based on the

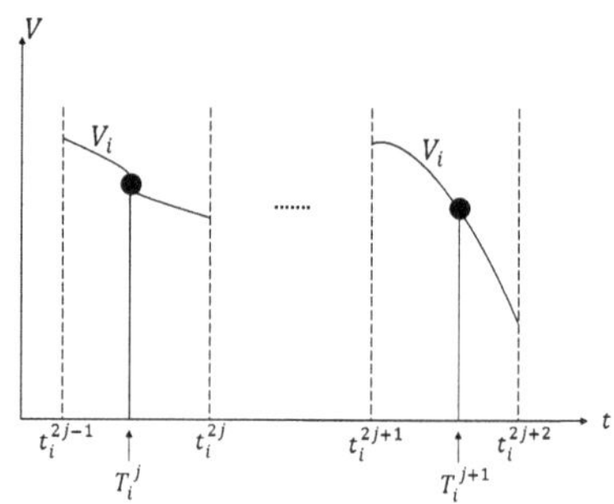

Fig. A.7 Illustration of the average values of $V_i(x(T_i^j))$

evaluation of the average value of Lyapunov functions during the intervals where the same subsystem is active.

Evidently, in the case where the subsystems are GAS, the result is practically equivalent to the previous results. However, his conditions are given with respect to the decrease of the average Lyapunov functions on the same intervals, see Fig. A.7.

Theorem A.8 [98] *Suppose that the N subsystems of (A.5), associated to N radially unbounded Lyapunov functions are GAS. Define the average value of the Lyapunov functions during the activation period for each subsystem as*

$$V_i\left(x\left(T_i^j\right)\right) \triangleq \frac{1}{t_i^{2j} - t_i^{2j-1}} \int_{t_i^{2j-1}}^{t_i^{2j}} V_i\left(x(\tau)\right) d\tau \quad \left(t_i^{2j-1} \leq T_i^j \leq t_i^{2j}\right) \qquad (A.11)$$

Then, the switched system is GAS in the sense of Lyapunov if, for all i,

$$V_i\left(x\left(T_i^{j+1}\right)\right) - V_i\left(x\left(T_i^j\right)\right) \leq -W_i\left(\left\|x\left(T_i^j\right)\right\|\right) \qquad (A.12)$$

holds for a positive definite continuous function $W_i(x)$.

Additionally, this result is extended to the case when the subsystems are not stable under the condition that the Lyapunov functions are bounded. In this case, if the average value of the Lyapunov functions decreases on the set of intervals associated to a subsystem i, then the switching system (A.5) is asymptotically stable, see Fig. A.8 [98].

Remark A.5 More recently, a similar result to the above using the average value of the derivative of the Lyapunov functions, rather than the average value of the Lyapunov functions, for the stability analysis of linear switching systems has been given by Michel in [54].

Recall that the stability is the first step in the study of a system in terms of its performance evaluation. In fact, if a system is not stable (or not stable enough), it is important to proceed to the stabilization of this system before looking to satisfy other performances such as trajectory tracking, precision, control effort, perturbation rejection, robustness, etc.

Fig. A.8 Illustration of the decrease of energy in the presence of unstable systems

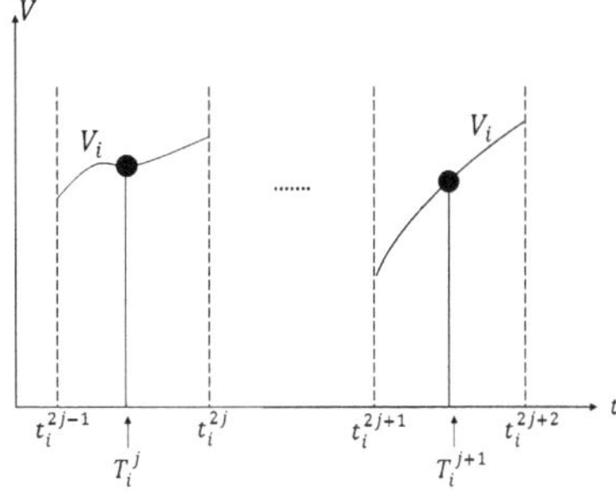

A.1.4 Stabilization of a System

The problem of stabilization consists in maintaining the system near an equilibrium point y^*. The aim is to construct stabilizing control laws such that y^* becomes an asymptotically stable equilibrium point of the system under these control laws.

Remark A.6

1. The problem of trajectory tracking consists in maintaining the solution of the system along a desired trajectory $y_d(t)$, $t \geq 0$. The objective here is to find a control law such that for every initial condition in a region D, the error between the output and the desired output

$$e(t) = y(t) - y_d(t)$$

   tends to 0 when $t \to \infty$. In addition, the state must remain bounded.
2. Note that the stabilization problem around an equilibrium point y^* is a special case of the problem of trajectory tracking whereby

$$y_d(t) = y^*, \quad t \geq 0$$

The control design techniques allowing to construct control laws for the stabilization of systems are numerous and varied. In what follows, we are going to present those that are most useful for the control of underactuated mechanical systems. The main references where most of the results on this subject were borrowed from, in the next section, are [32, 37, 41, 43, 44, 66, 67].

A.2 Control Theory

Given a physical system that we want to control and the system behavior we want to obtain, designing a control amounts to construct control laws such that the system subjected to these laws (the closed-loop system) presents the desired behavior.

Nonetheless, the control procedure is only possible if the system in question is controllable. Otherwise, the uncontrollable modes would need to be stable [13]. For more details, please refer to Appendix D.

The synthesis of control laws for nonlinear systems is difficult in general. Thereafter, we propose some control design techniques for the class of nonlinear control affine systems of the form

$$\dot{x} = f(x) + g(x)u, \quad x \in \mathbb{R}^n, \ u \in \mathbb{R} \tag{A.13}$$

The linearizability is a property that renders the systems more easy to control. In addition, the control design techniques for linear systems are well-established and largely developed. We can cite some examples such as pole placement control, optimal control, and a frequency-based approach just to mention a few. For more details on these subjects the interested reader can refer to [2, 17, 24, 35, 42, 56, 93]. This list is far from complete obviously.

Thus, it might be useful to highlight this linearizability property for nonlinear systems too. In what follows, the most employed and well-known procedures are briefly recalled.

A.2.1 Local Stabilization

Consider the system of the following form:

$$\dot{x} = f(x) + g(x)u, \quad f(0,0) = 0$$

In the presence of the control input u, the linear approximation around the equilibrium point is given by

$$\dot{x} = Ax + Bu \tag{A.14}$$

where the matrices A and B are defined by

$$A = \frac{\partial f}{\partial x}(0,0), \qquad B = \frac{\partial f}{\partial u}(0,0).$$

The obtained form (A.14) justifies the utilization of linear control techniques mentioned above.

Unfortunately, the resulting linearized system is typically valid only around the considered point so that the associated controller is valid only in a neighborhood of this point. This leads to a local control only. In addition, determining the linearity domain is not obvious.

In Appendix B, the reader will find more details of the limits of linearization and the underlying dangers of destabilization.

Hence, even though this method is simple and practical, it is necessary to proceed differently in order to increase the validity domain of the synthesized controllers.

To further benefit from the theory of linear control, there exists a control design technique based on a change of coordinates and a state feedback allowing to render the nonlinear dynamics equivalent to that of a linear dynamics: this is the so-called feedback linearization.

A.2.2 Feedback Linearization

When we transform a system via a change of coordinates, some of its properties remain unchanged. For example, if a system is unstable then the transformed system is also unstable. If a system is controllable, then the transformed system is also controllable. On the other hand, some systems might seem nonlinear in certain coordinates while they can become linear in other coordinates and under certain feedback.

Thus, it is interesting, whenever possible, to analyze the dynamics of a system in a transformed form that is easier to study.

In Appendix C, some results and concepts of differential geometry necessary for the presentation of this approach are recalled.

Two procedures of linearization by feedback are possible: input–state linearization and input–output linearization.

A.2.2.1 Input–State Linearization

The aim here is to transform the system of the form (A.13) via a diffeomorphism $z = \varphi(x)$ into a system of the form

$$
\begin{aligned}
\dot{z}_1 &= z_2 \\
\dot{z}_2 &= z_3 \\
&\ \vdots \\
\dot{z}_{n-1} &= z_n \\
\dot{z}_n &= a\big(\varphi^{-1}(z)\big) + b\big(\varphi^{-1}(z)\big)u
\end{aligned}
\tag{A.15}
$$

This form is similar to the canonical form of Brunovsky or the canonical form of controllability of linear systems.

If such transformation is possible, then for $b(\varphi^{-1}(z)) \neq 0$ the control

$$
u = \frac{1}{b(\varphi^{-1}(z))}\big(v - a\big(\varphi^{-1}(z)\big)\big)
\tag{A.16}
$$

permits to linearize the system, which becomes

$$
\dot{z}_1 = z_2, \qquad \dot{z}_2 = z_3, \qquad \ldots, \qquad \dot{z}_{n-1} = z_n, \qquad \dot{z}_n = v
$$

where v is an external control.

One can then ask the following questions: Is it always possible to linearize a system by feedback? When this is the case, how do we obtain the transformation $z = \varphi(x)$?

The answer to these questions lies in the following theorem:

Theorem A.9 [33] *The system* (A.13) *is input–state linearizable in a domain D if and only if:*

1. *The rank of the controllability matrix* $C_{fg} = \{g, ad_{fg}, \ldots, ad_{fg}^{n-1}\}$ *is equal to* n *for all* $x \in D$.
2. *The distribution* $\{g, ad_{fg}, \ldots, ad_{fg}^{n-2}\}$ *is involutive in* D.

With regard to the diffeomorphism, when the conditions of linearization are satisfied, then there exist several algorithms that permit to find the latter [14, 37, 65].

A.2.2.2 Input–Output Linearization

Consider the following nonlinear system:

$$\begin{aligned} \dot{x} &= f(x) + g(x)u, \quad x \in \mathbb{R}^n, \; u, y \in \mathbb{R} \\ y &= h(x) \end{aligned} \tag{A.17}$$

The idea is to generate linear equations between the output y and a certain input v through a diffeomorphism $z = \phi(x)$ constituted of the output and its derivatives with respect to time up to the order $n - 1$ when the relative degree r associated to this system is equal to n:

$$\begin{aligned} \phi_1(x) &= h(x) \\ \phi_2(x) &= L_f h(x) \\ &\;\;\vdots \\ \phi_n(x) &= L_f^{n-1} h(x) \end{aligned} \tag{A.18}$$

The system thus transformed is written as

$$\begin{aligned} \dot{z}_1 &= z_2 \\ \dot{z}_2 &= z_3 \\ &\;\;\vdots \\ \dot{z}_{n-1} &= z_n \\ \dot{z}_n &= a\big(\phi^{-1}(z)\big) + b\big(\phi^{-1}(z)\big)u. \end{aligned} \tag{A.19}$$

By choosing u of the form (A.16) and assuming that $b(\phi^{-1}(z)) \neq 0$ for all $z \in R^n$, the system becomes

$$\dot{z}_1 = z_2, \qquad \dot{z}_2 = z_3, \qquad \ldots, \qquad \dot{z}_{n-1} = z_n, \qquad \dot{z}_n = v.$$

Note that in this case, the form (A.19) is the same as in (A.15). In fact, when $r = n$ the two linearizations are equivalent. Hence, the conditions for applying the second linearization will be the same as for the first.

For more details on these two linearizations, and for some useful examples, the reader can refer to [32, 33, 37, 66].

Obviously, for a relative degree $r < n$, the system is no longer completely feed-back linearizable. In this case, one can talk of a partial feedback linearization.

A.2.2.3 Partial Feedback Linearization

When $r < n$, it is only possible to partially linearize a system of the form (A.17) through the diffeomorphism constituted partly by the output $h(x)$ and its successive derivatives up to order $r - 1$: $z = \phi_i(x)$ for $1 < i < r$, and completed—by using the theorem of Frobenius—by $n - r$ other functions: $\eta = \phi_i(x)$ for $r + 1 \leq i \leq n$, chosen in such a way that $L_g \phi_i = 0$ for $r + 1 \leq i \leq n$. In the coordinates (z, η) the equations of the system are given by

$$
\begin{aligned}
\dot{z}_1 &= z_2 \\
\dot{z}_2 &= z_3 \\
&\;\;\vdots \\
\dot{z}_{r-1} &= z_r \\
\dot{z}_r &= a(z, \eta) + b(z, \eta)u \\
\dot{\eta} &= q(z, \eta) \\
y &= z_1
\end{aligned}
\tag{A.20}
$$

This particular form is called the normal form.

If $b(z, \eta) \neq 0$, the input u can be chosen as

$$u = \frac{1}{b(z, \eta)} \big(v - a(z, \eta)\big).$$

In this case, the system takes the form

$$
\begin{aligned}
\dot{z}_1 &= z_2 \\
\dot{z}_2 &= z_3 \\
&\;\;\vdots \\
\dot{z}_{r-1} &= z_r
\end{aligned}
\tag{A.21}
$$

$$\dot{z}_r = v$$

$$\dot{\eta} = q(z, \eta)$$

Clearly, this system is composed of a linear subsystem of dimension r that is controllable by v—which is responsible for the input–output behavior—and of a nonlinear subsystem of dimension $n - r$ whose behavior is not affected by the control input. It follows that the global behavior of the system depends on this internal dynamics and that the verification of its stability is an essential step.

In [32], it was shown that the stability study of the internal dynamics can be reduced to that of the zero dynamics. This is obtained when we apply a control u that brings and maintains the output y to zero. In other words, the zero dynamics is given by the system $\dot{\eta} = q(0, \eta)$.

Remark A.7

1. When $\dot{\eta} = q(0, \eta)$ is (locally) asymptotically stable then the associated system is said to have (locally) minimum phase characteristic at the equilibrium point $\bar{x}$.
2. When $\dot{\eta} = q(0, \eta)$ is unstable then the associated system is said to be a non-minimum phase system.

Even though the methods of linearization are useful for simplifying the study and the control of nonlinear systems, they nevertheless present certain limitations. For example, the lack of robustness in the presence of modeling errors, the verification of certain conditions such as the involutivity, which, very often, is not verified by many systems; even those belonging to the class of nonlinear control affine systems, this is the case of UMSs. In addition, the state must be fully measured and accessible. Hence, the utilization of such techniques is confined to some classes of systems only.

Therefore, one must find other linearization techniques that are applicable to a wide range of systems without demanding restrictive and rigorous conditions as required by exact linearization. For example, approximative linearization techniques allow the linearization of the systems up to certain order and neglect certain nonlinear dynamics of high order. The authors that were interested in this technique are [5, 28, 31, 36, 39, 73], just to mention a few of them.

A.2.2.4 Approximate Feedback Linearization

For certain systems the computation of the relative degree presents some singularities in the neighborhood of the equilibrium point. For other systems the relative degree is smaller than the order of the system. In this case, the condition of involutivity is not verified.

The key idea of approximate linearization is to find an output function such that the system approximatively verifies the former condition.

Several linearization algorithms are available; one can cite, for example, linearization by the Jacobian, pseudo-linearization, Krener algorithm, Hunt and

Turi [39], the algorithm of Krener based on the theory of Poincaré [40], the algorithm of Hauser, and that of Sastry and Kokotović [28]. A comparative study of these algorithms applied to some examples can be found in [45].

In what follows, we shall recall briefly the algorithm of Hauser et al. [28]. Consider the system of the form (A.17):

$$\dot{x} = f(x) + g(x)u$$
$$y = h(x)$$

Suppose that the relative degree associated to this system is equal to $r < n$. Consequently, the system is not exactly feedback linearizable. In f and g, some terms prevent the linearization to take place, in the sense that the relative degree in the presence of these terms is smaller than n.

The idea is to neglect these terms so that we can achieve a complete relative degree, called robust relative degree.

Definition A.5 [88]   The robust relative degree of a regular output associated to system (A.17) at 0 is the integer γ such that

$$L_g h(0) = L_g L_f h(0) = \cdots = L_g L_f^{\gamma-2} h(0) = 0$$
$$L_g L_f^{\gamma-1} h(0) \neq 0$$

In this case, we say that the system (A.17) is approximatively feedback linearizable around the origin if there exists a regular output $y = h(x)$ for which $\gamma = n$.

This transformation is possible via the following diffeomorphism $z = \phi(x)$:

$$z_1 = h(x)$$
$$z_2 = L_f h(x)$$
$$\vdots$$
$$z_n = L_f^{n-1} h(x)$$

Hence, during an approximate linearization, the nonlinear model (A.17) is simplified by assuming that the functions $L_g h(x) = L_g L_f h(x) = \cdots = L_g L_f^{\gamma-2} h(x)$, preventing the definition of the classical relative degree and that cancel at 0, are identically zero:

$$L_g h(x) = L_g L_f h(x) = \cdots = L_g L_f^{\gamma-2} h(x) = 0$$

In this case, the system (A.17) is approximated by the following form:

$$\dot{z}_1 = z_2$$
$$\dot{z}_2 = z_3$$

$$\vdots \tag{A.22}$$

$$\dot{z}_{n-1} = z_n$$
$$\dot{z}_n = L_f^n h\big(\phi^{-1}(z)\big) + L_g L_f^{n-1} h\big(\phi^{-1}(z)\big)u$$

which is of the canonical form of Brunovsky.

Hence, if u is conveniently chosen (of type (A.16)) then, for $L_g L_f^{n-1} h(\phi^{-1}(z) \neq 0)$, the model will be in the linear form and will be locally controllable,

$$\dot{z}_1 = z_2, \qquad \dot{z}_2 = z_3 \qquad \dots, \qquad \dot{z}_{n-1} = z_n, \qquad \dot{z}_n = v$$

This method is in many cases satisfactory but naturally the control engineer will always try to improve it in order to increase its performances and its domain of validity. This is how the theory of higher order approximations was introduced by Krener [39] and Hauser [27].

A.2.2.5 Higher Order Approximations

The objective here is to maximize the order of terms to be neglected in order to have better precision. Hence, fewer terms are neglected. Consequently, the higher the order of the neglected residue is, the more effective the controller will be and the larger its domain of validity will be [8].

Theorem A.10 [39]   *The nonlinear system (A.13) can be approximated by a state feedback around an equilibrium point if and only if the distribution $\Delta_{n-2}(f, g)$ is involutive up to order*[3] $p - 2$ *on E.*

This means that there is a change of coordinates $z = \Psi(x)$ such that, in the new coordinates z, the dynamics (A.13) is given by

$$\begin{aligned} \dot{z}_1 &= z_2 \\ \dot{z}_2 &= z_3 \\ &\;\;\vdots \qquad\qquad\qquad +O_E^p(x) + O_E^{p-1}(x)u \\ \dot{z}_{n-1} &= z_n \\ \dot{z}_n &= a(z) + b(z)u \end{aligned} \tag{A.23}$$

with $b(\Psi^{-1}(x)) \neq 0$ in a neighborhood of E.

In other words, for the system (A.13), during a higher order approximation, the terms $L_g h(x) = L_g L_f h(x), \dots, L_g L_f^{\gamma-2} h(x)$ are no longer assumed to be zero but will be taken into account in the model and consequently in the expression of the control law.

---

[3]A distribution is involutive up to order p on E if $\forall f, g \in \Delta, [f, g] \in \Delta + O_E(\pi x)$.

The obtained model (A.23) is no longer fully linearizable, but it is at least easier to control than the initial system (A.13).

Apart from these methods of linearization, there exist several other approaches that are different from one another for the synthesis of control. The utilization of one method over another will depend on the class of systems considered. Among these methods, we shall be interested in three of them, namely: passivity approach, backstepping, and sliding mode control.

A.2.3 Few Words on Passivity

The notion of passivity is essentially linked to the notion of the energy that is accumulated in the considered system and the energy brought by external sources to the system [57, 67, 86]. The principal reference on the utilization of this concept of passivity in automatic control is due to Popov [61]. The dissipativity, which is an extension of this concept, is developed in the works of Willems [89].

Even though the concept of passivity is applicable to a large class of nonlinear systems, we will restrict our attention, only to dynamics modeled by system (A.17).

A dissipative system is then defined as follows:

Definition A.6 [67]   The system (A.17) is said to be dissipative if there exists a function $S(x)$ that is positive and such that $S(0) = 0$, and a function $w(u, y)$ that is locally integrable for all u, such that the following condition:

$$S(x) - S(x_0) \leq \int_t^0 w\big(u(\tau), y(\tau)\big) \, d\tau \tag{A.24}$$

is satisfied over the interval $[0, t]$.

This inequality expresses the fact that the energy stored in the system $S(x)$ is at most equal to the sum of energies initially stored and externally supplied. That is, there is no creation of internal energy; only a dissipation of energy is possible.

If $S(x)$ is differentiable, the expression (A.24) can be written as

$$\dot{S}(x) \leq w(u, y). \tag{A.25}$$

One particularity form of w permits to define the passivity of a system.

Definition A.7 [67]   The system (A.17) is said to be passive if it is dissipative and if the function w is a bilinear function from the input to the output $w(u, y) = u^T y$.

The passivity is a fundamental property of physical systems that is intimately linked to the phenomenon of energy loss or dissipation. One can recognize the principle similar to that of stability. In effect, the relation between passivity and stability can be established by considering the storage function $S(x)$ as a Lyapunov function $V(x)$.

Remark A.8 Note that the definition of dissipativity and passivity does not require that $S(x) > 0$ (it suffices that $S(x) \geq 0$). Hence, in the presence of an unobservable part, $x = 0$ can be unstable while the system is passive. For passivity to imply stability, one must exclude a similar case. That is, one must verify that the unobservable part is asymptotically stable. The reader should refer to [67] for a complete review on the stability of passive systems and for some results on Lyapunov functions that are semi positive definite.

A.2.4 Backstepping Technique

The backstepping is a recursive procedure for the construction of nonlinear control laws and Lyapunov functions that guarantee the stability of the latter. This technique is only applicable to a certain class of system which is said to be in strict feedback form (lower triangular). A quick review of this control design approach is given below, see [41] for more details.

Consider the problem of the stabilization of nonlinear systems in the following triangular form:

$$\dot{x}_1 = x_2 + f_1(x_1)$$
$$\dot{x}_2 = x_3 + f_2(x_1, x_2)$$
$$\vdots$$
$$\dot{x}_i = x_{i+1} + f_i(x_1, x_2, \ldots, x_i)$$
$$\vdots$$
$$\dot{x}_n = f_n(x_1, , x_2, \ldots, x_n) + u$$

(A.26)

The idea behind the backstepping technique is to consider the state x_2 as a "virtual control " for x_1. Therefore, if it is possible to realize $x_2 = -x_1 - f_1(x_1)$, then the state x_1 will be stabilized. This can be verified by considering the Lyapunov function $V_1 = \frac{1}{2}x_1^2$. However, since x_2 is not the real control for x_1, we make the following change of variables:

$$z_1 = x_1$$
$$z_2 = x_2 - \alpha_1(x_1)$$

with $\alpha_1(x_1) = -x_1 - f_1(x_1)$. By introducing the Lyapunov function $V_1(z_1) = \frac{1}{2}z_1^2$, we obtain

$$\dot{z}_1 = -z_1 + z_2$$
$$\dot{z}_2 = x_3 + f_2(x_1, x_2) - \frac{\partial \alpha_1}{\partial x_1}(x_2 + f_1(x_1)) := x_3 + \bar{f}_2(z_1, z_2)$$
$$\dot{V}_1 = -z_1^2 + z_1 z_2$$

By proceeding recursively, we define the following variables:

$$z_3 = x_3 - \alpha_2(z_1, z_2)$$

$$V_2 = V_1 + \frac{1}{2}z_2^2$$

In order to determine the expression of $\alpha_2(z_1, z_2)$, one can observe that

$$\dot{z}_2 = z_3 + \alpha_2(z_1, z_2) + \bar{f}_2(z_1, z_2)$$

$$\dot{V}_2 = -z_1^2 + z_2\big(z_1 + z_3 + \alpha_2(z_1, z_2) + \bar{f}_2(z_1, z_2)\big)$$

By choosing $\alpha_2(z_1, z_2) = -z_1 - z_2 - \bar{f}_2(z_1, z_2)$, we obtain

$$\dot{z}_1 = -z_1 + z_2$$

$$\dot{z}_2 = -z_1 - z_2 + z_3$$

$$\dot{V}_2 = -z_1^2 - z_2^2 + z_2 z_3$$

Proceeding recursively, at step i, and defining

$$z_{i+1} = x_{i+1} - \alpha_i(z_1, \ldots, z_i)$$

$$V_i = \frac{1}{2}\sum_{k=1}^{i} z_k^2$$

we obtain

$$\dot{z}_i = z_{i+1} + \alpha_i(z_1, \ldots, z_i) + \bar{f}_i(z_1, \ldots, z_i)$$

$$\dot{V}_i = -\sum_{k=1}^{i-1} z_k^2 + z_{i-1}z_i + z_i\big(z_{i+1} + \alpha_i(z_1, \ldots, z_i) + \bar{f}_i(z_1, \ldots, z_i)\big)$$

By using the expression $\alpha_i(z_1, \ldots, z_i) = -z_{i-1} - z_i - \bar{f}_i(z_1, \ldots, z_i)$, we obtain

$$\dot{z}_i = -z_{i-1} - z_i + z_{i+1}$$

$$\dot{V}_i = -\sum_{k=1}^{i-1} z_k^2 + z_i z_{i-1}$$

At step n, we obtain

$$\dot{z}_n = \bar{f}_n(z_1, \ldots, z_n) + u$$

Choosing

$$u = \alpha_n(z_1, \ldots, z_n) = -z_{n-1} - z_n - \bar{f}_n(z_1, \ldots, z_n)$$

for the following Lyapunov function:

$$V_n = \frac{1}{2} \sum_{k=1}^{n} z_k^2$$

it turns out that

$$\dot{z}_n = -z_{n-1} - z_n$$

$$\dot{V}_n = -\sum_{k=1}^{n} z_k^2$$

The stability of the system is proven by using simple quadratic Lyapunov functions. One must also note that the dynamic obtained in the z coordinates is linear. The advantage of the backstepping technique is its flexibility for the choice of the stabilizing functions α_i, which are simply chosen to eliminate all the nonlinearities in order to render the function $\dot{V}_i$ negative.

A.2.5 Sliding Mode Control

The theory of variable structure systems has been the subject of numerous studies over the last 50 years. Initial works on this type of systems are those of Anosov [1], Tzypkin [82]. and Emel'yanov [21]. These works have encountered a significant revival in the late 1970s when Utkin introduced the theory of sliding modes [83]. This control and observation technique received increasing interest because of their relative ease of design, their strength vis-à-vis certain parametric uncertainties and perturbations, and the wide range of their applications in varied fields such as robotics, mechanics or power systems.

The principle of this technique is to force the system to reach and then to remain on a given surface called sliding or switching surface (representing a set of static relationships between the state variables). The resulting dynamic behavior, called ideal sliding regime/mode, is completely determined by the parameters and equations defining the surface. The advantage of obtaining such behavior is twofold: on one hand, there is a reduction of the system order, and on the other, the sliding mode is insensitive to disturbances occurring in the same direction as the inputs.

The realization is done in two stages: a surface is determined so that the sliding mode has the desired properties (not necessarily present in the original system), and then a discontinuous control law is synthesized in order to make the surface invariant (at least locally) and attractive. However, the introduction of this discontinuous action, acting on the first derivative with respect to time of the sliding variable, does not generate an ideal sliding mode. On average, the controlled variables can be considered as ideally moving on the surface. In reality, the movement is characterized by high-frequency oscillations in the vicinity of the surface. This phenomenon is known as chattering and is one of the major drawbacks of this technique. Furthermore, it may stimulate non-modeled dynamics and lead to instability [23].

The presentation of this theory and its applications would easily constitute another book in itself. Therefore, in what follows, we swiftly present this technique and we refer the reader to [8, 22, 23, 60, 72, 83] for an excellent presentation of first order sliding modes and of the Fillipov theory for differential equations with discontinuous second member as well as of the equivalent vector method of Utkin.

A.2.5.1 Sliding Modes of Order One

Even though the theory of sliding modes is applied to a large class of systems of the form $\dot{x} = f(x, u)$ [69], we shall restrict our attention to the class of single-output control affine systems of the form

$$\dot{x} = f(x) + g(x)u \qquad (A.27)$$

where $x = (x_1, \ldots, x_n)^T$ belongs to χ, an open set of $\mathbb{R}^n$, u is the input and f, g are sufficiently differentiable functions. We define a sufficiently differentiable function $s : \chi \times \mathbb{R}^+ \to \mathbb{R}$ such that $\frac{\partial s}{\partial x}$ is non-zero on χ. The set

$$S = \left\{ x \in \chi : s(t, x) = 0 \right\} \qquad (A.28)$$

is a submanifold of χ of dimension $(n - 1)$, called the sliding surface. The function s is called the sliding function.

Remark A.9 The systems studied here are governed by differential equations involving discontinuous terms. The classical theories do not allow to describe the behavior of the solution in this case. One must, therefore, employ the theory of differential inclusions [3] and the solutions in the Fillipov sense [22].

Definition A.8 [84] We say that there exists an ideal sliding mode on S if there exists a finite time t_s such that the solution of (A.27) satisfies $s(t, x) = 0$ for all $t \geq t_s$.

The existence of the sliding mode is guaranteed by sufficient conditions: the sliding surface must be locally attractive, which can be mathematically translated as

$$\lim_{s \to 0^+} \frac{\partial s}{\partial x}(f + gu) < 0 \quad \text{and} \quad \lim_{s \to 0^-} \frac{\partial s}{\partial x}(f + gu) > 0 \qquad (A.29)$$

This condition translates the fact that, in a neighborhood of the sliding surface, the velocity vectors of the trajectories of the system must point towards this surface, see Fig. A.9 [8].

Hence, once the surface is intersected, the trajectories stay in an ε-neighborhood of S, and we say that the sliding mode is ideal if we have exactly $s(t, x) = 0$.

Fig. A.9 Attractivity of the surface

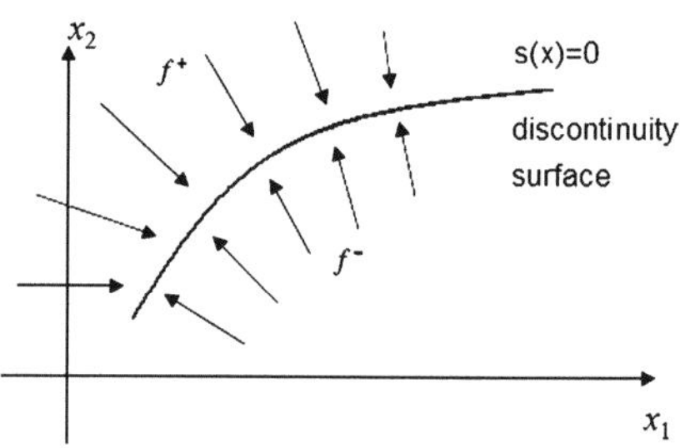

The condition (A.29) is often written in the form

$$s\dot{s} < 0 \tag{A.30}$$

and is called the attractivity condition.

The control u is constructed such that the trajectories of the system are brought towards the sliding surface and are then maintained in a neighborhood of the latter. u is a variable control law defined as follows:

$$u = \begin{cases} u^+(x) & \text{if } s(t,x) > 0, \\ u^-(x) & \text{if } s(t,x) < 0, \end{cases} \quad u^+ \neq u^- \tag{A.31}$$

with u^+ and u^- being continuous functions. It must be noted that it is this discontinuous characteristic of the control law that permits to obtain a convergence in finite time on the surface as well as the properties of robustness with respect to certain perturbations.

An example of a classical control by sliding mode that ensures the convergence towards the surface $s = 0$ in finite time is as follows: if for the nonlinear system (A.13) of relative degree r, we have $|L_g L_f^{r-1}| > K > 0$, $L_f^r < M < \infty$ then there exists $\lambda > 0$ such that the control [4]

$$u = -\operatorname{sign}\left(L_g L_f^{r-1}\right) \lambda \operatorname{sign}(s) \tag{A.32}$$

ensures the convergence of s to 0 in finite time.

Remark A.10 Often, we assume that $L_g h$ is positive. In this case, it is sufficient to take

$$u = -\lambda \operatorname{sign}(s). \tag{A.33}$$

Whenever that is not the case, it is more accurate to consider the expression (A.32).

A.2.5.2 Convergence in Finite Time

When on one hand, the control is chosen of the form (A.32) or simply of the form (A.33), and on the other hand, the previous conditions for the boundedness of certain functions are verified then the convergence in finite time is ensured. We shall try to demonstrate this result through an example.

Example A.1 [4] Consider the following simple example:

$$\dot{x} = b + u$$
$$u = -\lambda \operatorname{sign}(x - x_d) \tag{A.34}$$

with x_d the desired state, $s = x - x_d$ the sliding surface and $\lambda > |b| + sup|\dot{x}_d|$, then x converges to x_d in finite time and remains on the surface $x = x_d$.

Proof

$$s = x - x_d$$
$$\dot{s} = b - \lambda \operatorname{sign}(s) - \dot{x}_d$$

Consider the Lyapunov function: $V = \frac{s^2}{2}$. In this case, we have

$$\dot{V} = s\left(b - \lambda \operatorname{sign}(s) - \dot{x}_d\right)$$

if $\lambda > |b| + sup|\dot{x}_d|$ then $\dot{V} < 0$.

Hence, the convergence is demonstrated. It now remains to show that the convergence is achieved in finite time.

Since $\dot{V} < 0$, there exists a constant $K > 0$ such that $\dot{V} < -K|s|$.

Now $V = \frac{s^2}{2} \Rightarrow |s| = \sqrt{2V}$ therefore $\dot{V} < -\sqrt{2}K\sqrt{V}$,

Set $K_1 = -\sqrt{2}K \Rightarrow \dot{V} < -K_1\sqrt{V}$, let us take the worst case where the maximum convergence time is the limit case: $\dot{V} = -K_1\sqrt{V}$

The solution of this equation gives

$$V^{-1/2}dV = -K_1 dt$$
$$2V^{1/2} = -K_1 t + V_0$$
$$V(t) = \left(\frac{-K_1 t + V_0}{2}\right)^2$$

The time from which $V(t) = 0$ corresponds to $t = \frac{V_0}{K_1}$, which is finite. $\square$

A.2.5.3 The Chattering Phenomenon

In practice, an ideal sliding mode does not exist since it will imply that the control can switch with an infinite frequency. There then occurs the problem of chattering which means that we no longer have $s(t, x) = 0$ but $\|s(t, x)\| < \Delta$ from $t > t_0$ where t_0 is the convergence time and Δ a constant representing the maximum variations along the ideal trajectory $s = 0$.

This maximum depends on the "slew rate "of the components intervening in the injection of the input u in the system, on wear, and on the sensitivity of actuator

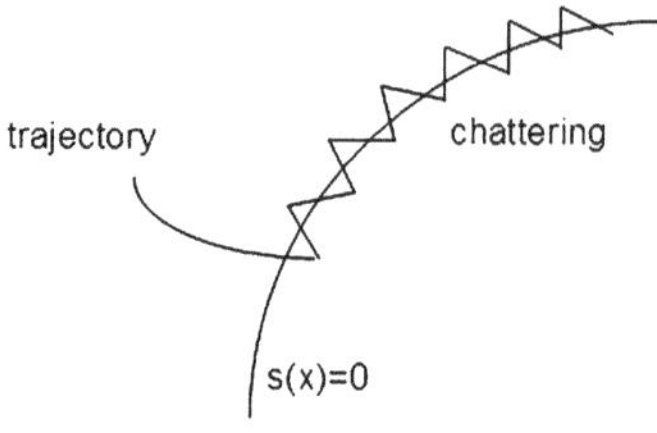

Fig. A.10 Chattering phenomenon

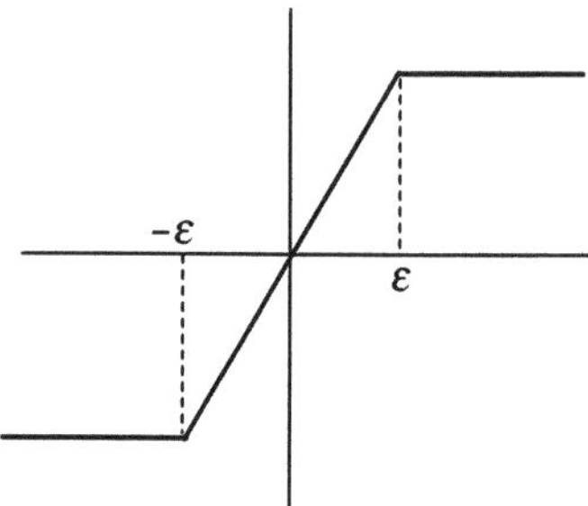

Fig. A.11 Saturation function

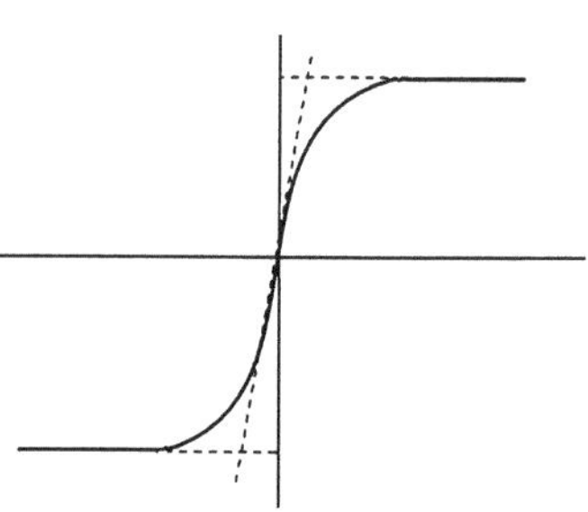

Fig. A.12 Sigmoid function

noise in the case of an analog control hence limiting the variation of speed between u^+ and u^-, see Fig. A.10. In the discrete case, the switching speed is limited by the data measurement which is in turn constrained by the sampling period and the computation time [8].

This phenomenon constitutes a non-negligible disadvantage because even if it is possible to filter the output of the process, it is susceptible to excite high-frequency modes that were not taken into account in the model of the system. This can degrade the performances and even lead to instability [29].

The chattering also implies important mechanical requirements at the actuator level. This can cause rapid wear and tear of the actuators as well as non-negligible energy loss at the power circuits level. Numerous studies have been carried out in order to reduce this phenomenon. One of them consists in replacing the sign function by saturation functions Fig. A.11, or by sigmoid functions such as $\tan(s)$ or $\arctan(s)$, Fig. A.12 [8].

Nevertheless, it has been proven that to overcome this chattering phenomenon the best solution is to consider higher order sliding modes such as the twisting algorithm or the super twisting [25, 60].

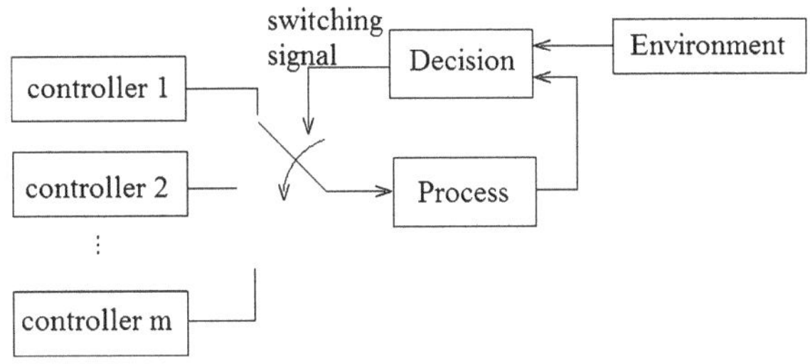

Fig. A.13 Architecture of multi controllers

A.2.6 Control Design Technique Based on the Switching Between Several Controllers

The control design techniques based on the switching between several controllers have been the subject of intensive applications these last few years. The importance of such methods comes from the existence of systems that are not stabilizable by a single controller. In effect, a large range of dynamical systems is modeled by a family of continuous subsystems and a logic rule orchestrating the switching between these subsystems, see Fig. A.13

Based on this, switching systems appear as a rigorous concept for studying complexes systems, even if their theoretical properties are still the subject of intensive research.

A.3 Summary

This appendix has been devoted to the presentation of the theoretical aspects on stability and control of nonlinear and switching systems. There is no general methodology for the design of controller for nonlinear systems as opposed to controller design for linear systems. Depending on the class of nonlinear systems under study, some approaches are better suited than others. In addition, we have attempted to explain in a simple way the principle of some nonlinear control design techniques that fall within the scope of this book with the aim of using some of them for the stabilization of underactuated mechanical systems.

Appendix B
Limits of Linearization and Dangers of Destabilization

A common practice of automatic community is to assume that a system can be described by a set of differential equations around an operating point as follows:

$$\dot{x} = Ax + Bu$$
$$y = Cx \tag{B.1}$$

Assuming that (B.1) describes the system behavior, we can then exploit linear control design procedures, where powerful analysis and design tools are available. However, nonlinear system behaviors can be more complex than what can be represented by an equivalent linear model.

Neglecting such behaviors, unpredictable instability may arise and may cause performance degradation. Moreover, the obtained linear system is valid only around the considered operating point. Hence, it can describe the system only in the neighborhood of this point. On the other hand, some phenomena such as Coulomb friction, backlash, and hysteresis, called hard nonlinearities, cannot be captured by linear equations. Therefore, these nonlinearities are neglected.

Additional nonlinear phenomena include finite escape time, multiple equilibrium points, limit cycles, and chaos. A more complete description of these phenomena and others is given in [18, 37].

To illustrate the impact of the loss of information due to linearization, let us consider the following examples [20]:

Example B.1   Several equilibrium points:

$$\dot{x} = -x + x^2$$
$$x(t = 0) = x_0 \tag{B.2}$$

After linearization around $x(t) = 0$, the obtained dynamic and associated solution are given by

$$\dot{x} = -x(t)$$
$$x(t) = x_0 e^{(-t)} \tag{B.3}$$

A. Choukchou-Braham et al., *Analysis and Control of Underactuated Mechanical Systems*, DOI 10.1007/978-3-319-02636-7,
© Springer International Publishing Switzerland 2014

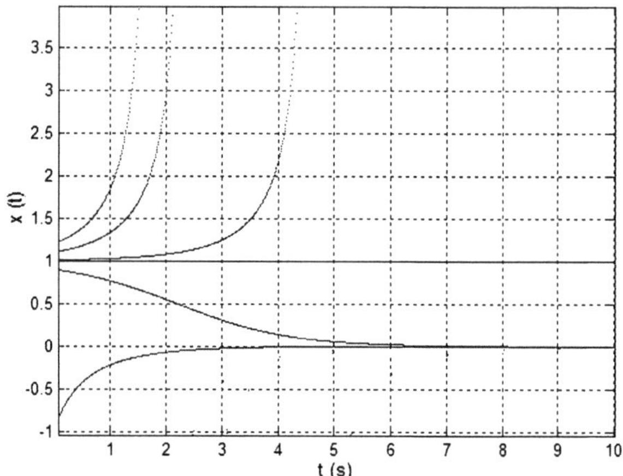

Fig. B.1 Responses of a nonlinear system for several initial conditions

Equation (B.3) indicates that for any initial condition x_0, the solution exponentially converges towards the equilibrium point.

However, according to (B.2), the nonlinear system possesses a second equilibrium point at $x(t) = 1$.

The impact of negligence of this point can be illustrated by calculating the solution of the nonlinear system:

$$x(t) = \frac{x_0 e^{-t}}{1 + x_0(e^{-t} - 1)} \tag{B.4}$$

according to (B.4), note that:

For $x_0 < 1$, the solution tends to 0 when $t \to \infty$ like for the linear case.

For $x_0 > 1$, the solution explodes to infinity in finite time, see Fig. B.1.

Example B.2 The linearized system is not controllable.

Consider the following unicycle robot model (see [7] for other models):

$$\begin{bmatrix} \dot{x}_1 \\ \dot{x}_2 \\ \dot{x}_3 \end{bmatrix} = \begin{bmatrix} \cos x_3 & 0 \\ \sin x_3 & 0 \\ 0 & 1 \end{bmatrix} \begin{bmatrix} u_1 \\ u_2 \end{bmatrix} \tag{B.5}$$

Clearly, (B.5) is controllable while the linearized system around the point $x_3(t) = 0$ given by

$$\begin{bmatrix} \dot{x}_1 \\ \dot{x}_2 \\ \dot{x}_3 \end{bmatrix} = \begin{bmatrix} 1 & 0 \\ 0 & 0 \\ 0 & 1 \end{bmatrix} \begin{bmatrix} u_1 \\ u_2 \end{bmatrix} \tag{B.6}$$

is not controllable for $x_2(t)$!

Other examples of performance degradation are given in [20].

On the other hand, the use of linear controller can sometimes lead to destabilization; for example, the consequence of the peaking phenomenon on a linear system can lead to the system instability [78, 80].

To illustrate this concept let us consider the partially linear coupled system described by the dynamics:

$$\dot{x} = f(x, y)$$
$$\dot{y} = Ay + By \tag{B.7}$$

Let us make the following assumptions:

(b1) The pair (A, B) is supposed controllable.
(b2) The nonlinear function f is differentiable to first order with respect to the time.
(b3) The origin is an GAS equilibrium point for the zero dynamic $\dot{x} = f(x, 0)$.

According to (B.7), and assumption (b3), it seems rather clear intuitively that a linear controller can be designed to lead the dynamics $y(t)$ to 0 exponentially such that the zero dynamic of the nonlinear system is GAS. However, this strategy can lead the nonlinear dynamics to instability and its trajectory may escape to infinity in finite time. For example consider the following system:

Example B.3 Finite escape time:

$$\dot{x} = \frac{-(1 + y_2)}{2} x^3 \tag{B.8}$$

$$\dot{y}_1 = y_2$$

$$\dot{y}_2 = u$$

From (B.8), one can verify that the assumption (b3) is satisfied.
By designing a linear controller as follows:

$$u = -a^2 y_1 - 2a y_2 \tag{B.9}$$

multiple eigenvalues for the closed system result at $-a$.

Linear analysis tools can be used to find the exact solution $y_2(t)$ given by

$$y_2 = -a^2 t e^{-at} \tag{B.10}$$

and from this solution it appears that the dynamic $|y_2(t)|$ rises to a peak, then converges exponentially to 0. By computation, we can show that the peak time is $t = \frac{1}{a}$.

From (B.10), we can conclude that for important values of a, y converges faster towards 0. Hence, from assumption (b3), it seems that important values of a allow a fast stabilization of the nonlinear system.

Nevertheless, in [38], it was shown that this is not true. Indeed, if we substitute (B.10) in (B.8). By integrating, the resulting expression is given by

$$x^2(t) = \frac{x_0^2}{1 + x_0^2(t + (1 + at)e^{-at} - 1)} \tag{B.11}$$

The peaking phenomenon destabilization effect is now apparent when we replace the values of x_0, a and t in (B.11). For example, for $a = 10$ and $x_0^2 = 2.176$, the response $x^2(t \cong 0.5)$ becomes unbounded and we have escape to infinity in finite time.

Other examples and discussion of this phenomenon are in [38, 78, 80].

Appendix C
A Little Differential Geometry

This section is devoted to the definition of some concepts and basic tools of differential geometry introduced in nonlinear automatic control theory, since the early 1970s, by Eliott, Lobry, Hermann, Krener, Brockett, and others.

Diffeomorphism A diffeomorphism is a nonlinear change of coordinates $z = \Phi(x)$ where Φ is a vectorial function

$$\Phi(x) = \begin{pmatrix} \Phi_1(x_1, \ldots, x_n) \\ \Phi_2(x_1, \ldots, x_n) \\ \vdots \\ \Phi_n(x_1, \ldots, x_n) \end{pmatrix}$$

with the following properties:

- $\Phi(x)$ is a bijective application
- $\Phi(x)$ and Φ^{-1} are differentiable applications.

If these properties are verified for all $x \in \mathbb{R}^n$ then Φ is a global diffeomorphism. Otherwise, Φ is a local diffeomorphism.

Proposition C.1 *If the Jacobian matrix of Φ, evaluated at the point $x = x_0$ is nonsingular, then $\Phi(x)$ is a local diffeomorphism.*

Lie Derivative and Bracket Let f and g be two vector fields on an open Ω of $\mathbb{R}^n$ with all continuous partial derivatives, and denote by $\frac{\partial f}{\partial x}$ and $\frac{\partial g}{\partial x}$ the Jacobian matrices.

The Lie derivative of g along f is the vector field

$$L_f g = \frac{\partial g}{\partial x} f.$$

The Lie bracket of f and g is the vector field

$$[f, g] = L_f g - L_g f.$$

A. Choukchou-Braham et al., *Analysis and Control of Underactuated Mechanical Systems*, DOI 10.1007/978-3-319-02636-7,
© Springer International Publishing Switzerland 2014

We define also the vector fields

$$ad f_f g = [f, g]$$

$$ad_f^k g = \left[f, ad_f^{k-1} g \right], \quad k = 2, 3, \ldots$$

Distribution, Involutivity and Complete Integrability

- A distribution Δ on a manifold M assigns to each point $x \in M$ a subspace of the tangent space T.
- A set of vectors $\{g_1, \ldots, g_m\}$ in Ω is said involutive if for all the i and j the bracket $[g_i, g_j]$ is a linear combination of vectors $g_1, \ldots, g_m$, that is, there exist functions α_{ij}^k defined in Ω such that

$$[g_i, g_j] = \sum_{k=1}^{k=m} \alpha_{ij}^k g_k.$$

  That is, if for all the f and g in Δ, $[f, g]$ belongs to Δ (Δ is closed by Lie bracket).
- A set of linearly independent vectors $\{g_1, \ldots, g_m\}$ is a complete integrable set, if the system of $n - m$ partial derivative equations

$$\frac{\partial h}{\partial x} g_1 = 0, \quad \ldots, \quad \frac{\partial h}{\partial x} g_{n-m} = 0$$

  admits a solution $h : \Omega \to \mathbb{R}^n$ such that $\frac{\partial h}{\partial x} \neq 0$

Theorem C.1 ([70] (Frobenius))   *A set of linearly independent vectors $\{g_1, \ldots, g_m\}$ is involutive if and only if it is completely integrable.*

For proof and examples see [37].

Relative Degree   The relative degree associated with the system

$$\begin{aligned} \dot{x} &= f(x) + g(x)u \\ y &= h(x) \end{aligned} \tag{C.1}$$

in a region $\Omega \subset \mathbb{R}^n$ is given by the integer γ such that

$$L_g h(x) = L_g L_f h(x) = \cdots = L_g L_f^{\gamma-2} h(x) = 0$$

$$L_g L_f^{\gamma-1} h(x) \neq 0$$

for all $x \in \Omega$.

Appendix D
Controllability of Continuous Systems

One of the main goals of automatic control is to establish control laws so that a system evolves according to a predetermined objective. This requires controllability of the system. Intuitively, the controllability means that we can bring a system from one state to another by means of a control. Conversely, non-controllability implies that some states are unreachable for any control.

D.1 Controllability of Linear Systems

For controlled linear systems

$$\dot{x} = Ax + Bu \tag{D.1}$$

$$y = Cx \tag{D.2}$$

where $A_{n \times n}$ is the state matrix, $x \in \mathbb{R}^n$ is the vector states, $B_{n \times m}$ the control matrix, u controls belonging to a set of admissible controls U, $C_{p \times n}$ the output matrix and $y \in \mathbb{R}^p$ the system outputs.

Definition D.1 [35]   The system (D.1) is controllable if for each couple (x_0, x_d) of $\mathbb{R}^n$ there exist a finite time T and a control u defined on $[0, T]$ that brings the system from an initial state $x(0) = x_0$ to a desired state $x(T) = x_d$.

D.1.1 Kalman Controllability Criterion

An algebraic characterization of linear systems controllability, due to Kalman, is given as follows:

A. Choukchou-Braham et al., *Analysis and Control of Underactuated Mechanical Systems*, DOI 10.1007/978-3-319-02636-7,
© Springer International Publishing Switzerland 2014

Fig. D.1 Static state feedback stabilization

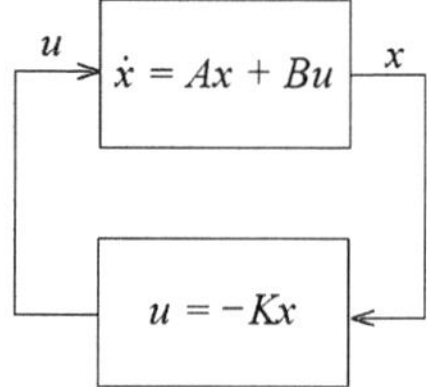

Theorem D.1 *The linear system* (D.1) *is controllable if and only if the rank of its controllability matrix*

$$\mathbf{C} = \begin{pmatrix} B & AB & \ldots & A^{n-1}B \end{pmatrix} \tag{D.3}$$

is equal to n. We say that the pair (A, B) *is controllable.*

Details and proofs can be found in [17].

For controllable linear systems, we seek to design a controller that makes the origin asymptotically stable. Several approaches are available, for example, we can design control laws by state feedback.

D.1.2 State Feedback Stabilization

A linear state feedback or controller for (D.1) is a control law

$$u(t) = -Kx(t) \tag{D.4}$$

where $K_{m \times n}$ is a gain matrix.

When the value of $u(t)$ at t depends only on $x(t)$ then the feedback is called static feedback, Fig. D.1.

The gain matrix can be computed in several ways, for example by pole placement.

Pole Placement Design When a system is controllable, the pole placement principle consists of determining a control law $u = -Kx$ such that $\sigma(A - BK) = \sigma_d$, where σ is the spectrum of $(A - BK)$ and σ_d is the desired spectrum.

The difficulty of this approach lies in the determination of the spectrum since there is no general methodology for doing so. This method offers the possibility to place the closed-loop poles anywhere in the negative half-plane, regardless of the open-loop poles location. As a result, the response time can be controlled. However, if the poles are placed too far into the negative half-plane, the values of the gain K are very large and can cause saturation problems and can lead to instability.

Remark D.1 The control law u is designed assuming that the state vector x is available. This assumption is not always true. Sometimes, some states are not available, because it is either difficult or impossible to physically measure these states or it is too expensive. In this case, we proceed to a reconstruction of the missing states using observers. However, throughout this book we are interested in the problem of control under the assumption that states are measurable.

D.2 Controllability Concepts for Nonlinear Systems

The notion of controllability which seems simple and intuitive for linear systems is rather complicated for nonlinear systems where several definitions of the latter exist. The first results on nonlinear system controllability are due to Sussmann and Jurdjevic [79], Lobry [48], Hermann and Krener [30], Sussmann [76, 77] and for a nice presentation see also Nijmeijer and Van der Schaft [55].

A nonlinear system is generally represented by

$$\dot{x} = f(x, u)$$
$$y = h(x) \tag{D.5}$$

where $x \in M \subset \mathbb{R}^n$, $u \in \mathbb{R}^m$, $y \in \mathbb{R}^p$, and f, h are C^∞.

Definition D.2 Let U be a subset of M and let $(x_0, x_d) \in U$. We say that x_d is U-accessible from x_0 which we denote by $x_d A_u x_0$, if there exist a measurable and bounded control u and a finite time T, such that the solution $x(t)$ of (D.5), for $t \in [0, T]$, satisfies

$$x(0) = x_0, \qquad x(T) = x_d \quad \text{and} \quad x \in U \quad \text{for } t \in [0, T]$$

we denote by $A(x_0)$ the set of points in M accessible from x_0:

$$A(x_0) = \{x \in M / \ x A_M x_0\} \tag{D.6}$$

Definition D.3 The system (D.5) is controllable at x_0 if $A(x_0) = M$ and is controllable if $A(x_0) = M$ for all $x \in M$.

When a system is controllable at x_0, it may be necessary to cover a considerable distance or time for reaching a point near x_0. This leads us to introduce a local version of the concept of controllability.

Definition D.4 The system (D.5) is said locally controllable at x_0, if for all neighborhood U of x_0, $A_u(x_0)$ is a neighborhood of x_0, where

$$A_u(x_0) = \{x \in U / \ x A_u x_0\} \tag{D.7}$$

and it is said to be locally controllable if it is locally controllable for all $x \in M$.

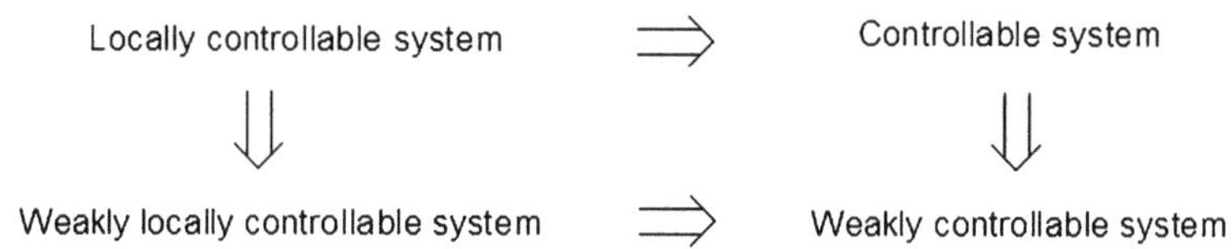

Fig. D.2   Relationship between the nonlinear controllability concepts

We can weaken the controllability notion:

Definition D.5   The system is said to be weakly controllable at x_0, if $WA(x_0) = M$, it is said to be weakly controllable if it is weakly controllable for all $x \in M$.

Remark D.2 WA_u is the smallest set containing U-accessible pairs (that is, $x^{'} WA_u x^{''}$ if and only if there exist $x^0, \ldots, x^k$, such that $x^0 = x^{'}$, $x^k = x^{''}$ and let $x^i A_u x^{i-1}$ or $x^{i-1} A_u x^i$ for $i = 1, \ldots, k$).

The concept of weak controllability is a global concept which does not reflect the behavior of a system in the neighborhood of a point. Therefore, it is necessary to introduce the concept of weak local controllability

Definition D.6   The system (D.5) is said to be locally weakly controllable at x_0 if for all neighborhood U of x_0, $WA_u(x_0)$ is a neighborhood of x_0 and it is said to be locally weakly controllable if it is for all $x \in M$.

Different notions of controllability are illustrated by the diagram of Fig. D.2. For control affine nonlinear systems described by

$$\dot{x} = f(x) + g(x)u$$
$$y = h(x) \tag{D.8}$$

the controllability rank condition is defined as

Definition D.7   System (D.8) satisfies the rank condition if the rank of the nonlinear controllability matrix

$$\mathbf{C}_{fg} = \begin{bmatrix} g(x) & ad_f g(x) & ad_f^2 g(x) & \ldots & ad_f^{n-1} g(x) \end{bmatrix} \tag{D.9}$$

is equal to n for all x.

Theorem D.2 [30]   *If the system (D.8) satisfies the rank condition then it is locally weakly controllable.*

This theorem highlights the advantage of weak local controllability compared to previous forms of controllability since verification of such a concept is reduced to a simple algebraic criterion.

References

1. D.V. Anosov, On stability of equilibrium points of relay systems. J. Autom. Remote Control **2**,135–149 (1959)
2. P.J. Antsalkis, A.N. Michel, *Linear Systems* (McGraw-Hill, New York, 1997)
3. J-P. Aubin, H. Frankowska, *Set-Valued Analysis* (Birkhäuser, Basel, 1990)
4. J.P. Barbot, Systèmes à structure variable. Technical report, École Nationale Supérieure d'Electronique et de Ses Applications, ENSEA, France, 2009
5. N. Bedrossian, Approximate feedback linearisation: the cart pole example, in *Proc. IEEE Int. Conf. on Robotics and Automation* (1992), pp. 1987–1992
6. I. Belgacem, Automates hybrides Zénon: Exécution-simulation. Master at Aboubekr Belkaid University, Tlemcen, Algeria, 2009
7. A.M. Bloch, M. Reyhanoglu, N.H. McClamroch, Control and stabilization of nonholonomic dynamic systems. IEEE Trans. Autom. Control **37**(11),1746–1757 (1992)
8. T. Boukhobza, Contribution aux formes d'observabilité pour les observateurs à modes glissants. Ph.D. thesis, Paris Sud Centre d'Orsay University, France, 1997
9. M.S. Branicky, Stability of switched and hybrid systems, in *Proc. 33rd IEEE Conf. on Decision and Control*, USA (1994), pp. 3498–3503
10. M.S. Branicky, Studies in hybrid systems: modeling, analysis, and control. Ph.D. thesis, Department of Electrical and Computer Engineering, Massachusetts Institute of Technology, 1995
11. M.S. Branicky, Stability of hybrid systems: State of the art, in *Proc. 36th IEEE Conf. on Decision and Control*, USA (1997), pp. 120–125
12. M.S. Branicky, Multiple Lyapunov functions and other analysis tools for switched and hybrid systems. IEEE Trans. Autom. Control **43**(4), 475–482 (1998)
13. R.W. Brockett, *Asymptotic Stability and Feedback Stabilization* (Birkhäuser, Basel, 1983)
14. K. Busawon, Lecture notes in control systems engineering. Technical report, Northumbria University, Newcastle, United Kingdom, 2008
15. S. Canudas, G. Bastin, *Theory of Robot Control* (Springer, Berlin, 1996)
16. G. Chesi, Y.S. Hung, Robust commutation time for switching polynomial systems, in *Proc. 46th IEEE Conf. on Decision and Control*, USA (2007)
17. B. D'Andréa Novel, M.C. De-Lara, *Commande Linéaire des Systèmes Dynamiques* (École des Mines Press, 2000)
18. H. Dang Vu, C. Delcarte, *Bifurcation et Chaos* (Ellipses edition, 2000)
19. R. DeCarlo, M. Branicky, S. Petterssond, B. Lennartson, Perspectives and results on the stability and stabilizability of hybrid systems. Proc. IEEE **88**, 1069–1082 (2000)
20. W.E. Dixon, A. Behal, D. Dawson, S. Nagarkatti, *Nonlinear Control of Engineering Systems* (Birkhäusser, Basel, 2003)
21. S.V. Emel'yanov, *Variable Structure Control Systems* (Nauka, Moscow, 1967)
22. A.G. Fillipov, *Differential Equations with Discontinuous Right Hand-Sides*. Mathematics and its Applications (Kluwer Academic, Dordrecht, 1983)
23. T. Floquet, Contributions à la commande par modes glissants d'ordre supérieur. Ph.D. thesis, University of Sciences and Technology of Lille, France, 2000
24. G.F. Franklin, J.D. Powell, A. Emami-Naeini. *Feedback Control of Dynamic Systems* (Prentice-Hall, Englewood Cliffs, 2002)
25. L. Fridman, A. Levant, *Sliding Modes of Higher Order as a Natural Phenomenon in Control Theory* (Springer, Berlin, 1996)
26. L. Gurvits, R. Shorten, O. Mason, On the stability of switched positive linear systems. IEEE Trans. Autom. Control **52**, 1099–1103 (2007)
27. J. Hauser, Nonlinear control via uniform system approximation. Syst. Control Lett. **17**, 145–154 (1991)

28. J. Hauser, S. Sastry, P. Kokotović, Nonlinear control via approximate input-output linearization. IEEE Trans. Autom. Control **37**(3), 392–398 (1992)
29. B. Heck, Sliding mode control for singulary perturbed systems. Int. J. Control **53**(4) (1991)
30. R. Hermann, A.J. Krener, Nonlinear controllability and observability. IEEE Trans. Autom. Control **22**, 728–740 (1977)
31. L.R. Hunt, R. Su, G. Meyer Gurvits, R. Shorten, O. Mason, Global transformations of nonlinear systems. IEEE Trans. Autom. Control **28**(1), 24–31 (1983)
32. A. Isidori, *Nonlinear Control Systems* (Springer, Berlin, 1995)
33. B. Jakubczyk, W. Respondek, On linearisation of control systems. Rep. Sci. Pol. Acad. **28**(9), 517–522 (1980)
34. J. Jouffroy, Stabilité et systèmes non linéaire: réflexion sur l'analyse de contraction. Ph.D. thesis, Savoie University, France, 2002
35. T. Kailath, *Linear Systems* (Prentice-Hall, Englewood Cliffs, 1981)
36. W. Kang, Approximate linearisation of nonlinear control systems, in *Proc. IEEE Conf. on Decision and Control* (1993), pp. 2766–1771
37. H.K. Khalil, *Nonlinear Systems* (Prentice-Hall, Englewood Cliffs, 2002)
38. P. Kokotović, The joy of feedback: Nonlinear and adaptive. IEEE Control Syst. Mag. **12**(3), 7–17 (1992)
39. A.J. Krener, Approximate linearisation by state feedback and coordinate change. Syst. Control Lett. **5**, 181–185 (1984)
40. A.J. Krener, M. Hubbard, S. Karaban, B. Maag, *Poincaré's Linearisation Method Applied to the Design of Nonlinear Compensators* (Springer, Berlin, 1991)
41. M. Kristić, P. Kokotović, *Nonlinear and Adaptive Control Design* (Wiley, New York, 1995)
42. H. Kwakernak, R. Sivan, *Linear Optimal Control Systems* (Library of Congress, 1972)
43. H.G. Kwatny, G.L. Blankenship, *Nonlinear Control and Analytical Mechanics: A Computational Approach* (Birkhäuser, Basel, 2000)
44. F. Lamnabhi Lagarrique, P. Rouchon, *Commandes Non Linéaires* (Lavoisier, Paris, 2003)
45. M. Latfaoui, Linéaisation approximative par feedback. Master at Aboubekr Belkaid University, Tlemcen, Algeria, 2002
46. J. Lévine, Analyse et commande des systèmes non linéaires. Technical report, École des Mines de Paris, France, 2004
47. D. Liberzon, A.S. Morse, Basic problems in stability and design of switched systems. IEEE Control Syst. Mag. **19**(5), 59–70 (1999)
48. C. Lobry, *Contrôlabilité des Systèmes Non Linéaires* (CNRS, Paris, 1981)
49. D.G. Luenberger, *Introduction to Dynamic Systems* (Wiley, New York, 1979)
50. N. Manamanni, Aperçu rapide sur les systèmes hybrides continus. Technical report, University of Reims, France, 2007
51. M. Margaliot, M.S. Branicky, Nice reachability for planar bilinear control systems with applications to planar linear switched systems. IEEE Trans. Autom. Control (May 2008)
52. O. Mason, R. Shorten, A conjecture on the existence of common quadratic Lyapunov functions for positive linear systems, in *Proc. American Control Conf.*, USA (2003), pp. 4469–4470
53. S. Mastellone, D.M. Stipanović, M.W. Spong, Stability and convergence for systems with switching equilibria, in *Proc. 46th IEEE Conf. on Decision and Control*, USA (2007), pp. 4013–4020
54. A.N. Michel, L. Hou, Stability results involving time-averaged Lyapunov function derivatives. J. Nonlinear Anal., Hybrid Syst. **3**(1), 51–64 (2009)
55. H. Nijmeijer, A. van der Schaft, *Nonlinear Dynamical Control Systems* (Springer, Berlin, 1990)
56. K. Ogata, *Modern Control Engineering* (Prentice-Hall, Englewood Cliffs, 1997)
57. R. Ortega, A. Loria, P. Nicklasson, H. Sira-Ramirez, *Passivity-Based Control of Euler Lagrange Systems* (Springer, Berlin, 1998)

58. P. Peleties, Modeling and design of interacting continuous-time/discrete event systems. Ph.D. thesis, Electrical Engineering, Purdue Univ., West Lafayette, IN, 1992
59. P. Peleties, R.A. DeCarlo, Asymptotic stability of m-switched systems using Lyapunov-like functions, in *Proc. American Control Conf.*, USA (1991), pp. 1679–1684
60. W. Perruquetti, J.P. Barbot, *Sliding Mode Control in Engineering* (Taylor and Francis, London, 2002)
61. V.M. Popov, Absolute stability of nonlinear control systems of automatic control. J. Autom. Remote Control **22**, 857–875 (1962)
62. J.P. Richard, *Mathématiques pour les Systèmes Dynamiques* (Lavoisier, Paris, 2002)
63. H. Saadaoui, Contribution à la synthèse d'observateurs non linéaires pour des classes de systèmes dynamiques hybrides. Ph.D. thesis, Cergy Pontoise University, France, 2007
64. R.G. Sanfelice, A.R. Teel, R. Sepulchre, A hybrid systems approach to trajectory tracking control for juggling systems, in *Proc. 46th IEEE Conf. on Decision and Control*, USA (2007), pp. 5282–5287
65. T. Sari, C. Lobry, *Contrôle Non Linéaire et Applications* (Hermann, Paris, 2005)
66. S.S. Sastry, *Nonlinear Systems: Analysis, Stability and Control* (Springer, Berlin, 1999)
67. R. Sepulchre, M. Janković, P. Kokotović, *Contructive Nonlinear Control* (Springer, Berlin, 1997)
68. C. Shen, Q. Wei, F. Shumin, On exponential stability of switched systems with delay: multiple Lyapunov functions approach, in *Proc. Chineese Control Conf.*, China (2007), pp. 664–668
69. H. Sira Ramirez, Sliding regimes in general nonlinear systems: a relative degree approach. Int. J. Control **50**(4), 1487–1506 (1989)
70. J. Slotine, W. Li, *Nonlinear Systems Analysis* (Prentice-Hall, Englewood Cliffs, 1993)
71. E.D. Sontag, Y. Wang, On characterizations of the input-to-state stability property. Syst. Control Lett. **24**, 351–359 (1995)
72. S.K. Spurgeon, C. Edwards, *Sliding Mode Control: Theory and Applications* (Taylor and Francis, London, 1983)
73. T. Sugie, K. Fujimoto, Control of inverted pendulum systems based on approximate linearisation, in *Proc. IEEE Conf. on Decision and Control* (1994), pp. 1647–1649
74. Z. Sun, S.S. Ge, Analysis and synthesis of switched linear control systems. Automatica **41**, 181–195 (2005)
75. Z. Sun, S.S. Ge, *Switched Linear Systems* (Springer, Berlin, 2005)
76. H.J. Sussmann, *Lie Brackets, Real Analyticity and Geometric Control* (Birkhäuser, Basel, 1983)
77. H.J. Sussmann, A general theorem on local controllability. SIAM J. Control Optim. **25**(1), 158–194 (1987)
78. H.J. Sussmann, Limitations on stabilizability of globally minimum phase systems. IEEE Trans. Autom. Control **35**(1), 117–119 (1990)
79. H.J. Sussmann, V. Jurdjevic, Controllability of nonlinear systems. J. Differ. Equ. **12**(2), 95–116 (1991)
80. H.J. Sussmann, P. Kokotović, The peaking phenomenon and the global stabilization of nonlinear systems. IEEE Trans. Autom. Control **36**(4), 424–440 (1991)
81. C.J. Tomlin, J. Lygeros, S. Sastry, *Introduction to Dynamic Systems* (Springer, Berlin, 2003)
82. Y.Z. Tzypkin, *Theory of Control of Relay Systems* (Gostekhizdat, Moscow, 1955)
83. V.I. Utkin, Variable structure systems with sliding mode. IEEE Trans. Autom. Control **22**(2), 212–222 (1977)
84. V.I. Utkin, *Sliding Modes in Control Optimization* (Springer, Berlin, 1992)
85. A. van der Schaft, H. Schumacher, *An Introduction to Hybrid Dynamical Systems* (Springer, Berlin, 2002)
86. A. van der Shaft, L_2 *Gain and Passivity Techniques in Nonlinear Control* (Springer, Berlin, 2000)

87. M. Vidyasagar, *Nonlinear Systems Analysis* (Prentice-Hall, Englewood Cliffs, 1993)
88. D.A. Voytsekhovsky, R.M. Hirschorn, Stabilization of single-input nonlinear systems using higher order compensating sliding mode control, in *Proc. 45th IEEE Conf. on Decision and Control and the European Control Conf.* (2005), pp. 566–571
89. J.C. Willems, Dissipative dynamical systems. Rational mechanics and analysis **45**, 321–393 (1972)
90. X. Xu, G. Zhai, Practical stability and stabilization of hybrid and switched systems. IEEE Trans. Autom. Control **50**(11), 1897–1903 (2005)
91. H. Ye, A.N. Michel, L. Hou, Stability analysis of discontinuous dynamical systems with application, in *13th IFAC Congress*, USA (1996), pp. 461–466
92. H. Ye, A.N. Michel, L. Hou, Stability theory for hybrid dynamical systems. IEEE Trans. Autom. Control **43**(4), 461–474 (1998)
93. S.H. Zak, *Systems and Control* (Oxford University Press, London, 2003)
94. G. Zames, On the input-output stability of nonlinear time-varying feedback systems. IEEE Trans. Autom. Control **11**(2), 228–238 (1966)
95. G. Zhai, B. Hu, K. Yasuda, A.N. Michel, Stability analysis of switched systems with stable and unstable subsystems: an average dwell time approach, in *Proc. American Control Conf.*, USA (2000), pp. 200–204
96. G. Zhai, D. Liu, J. Imae, T. Kobayashi, Lie algebraic stability analysis for switched systems with continuous-time and discrete-time subsystems. IEEE Trans. Circ. Syst. **53**(2), 152–156 (2006)
97. G. Zhai, H. Lin, X. Xu, A.N. Michel, Stability analysis and design of switched of normal systems, in *Proc. 43rd IEEE Conf. on Decision and Control*, USA (2004), pp. 3253–3258
98. G. Zhai, I. Matsune, J. Imae, T. Kobayashi, A note on multiple Lyapunov functions and stability condition for switched and hybrid systems, in *Proc. 16th IEEE Int. Conf. on Control Application*, Singapore (2007), pp. 1189–1199
99. J. Zhao, D.J. Hill, Dissipativity theory for switched systems. IEEE Trans. Autom. Control **53**(4), 941–953 (2008)

Index

A
Acrobot, 8, 9, 18, 27, 29, 42, 47, 64, 65
Approximation
    high order, 83, 85, 112
    linear, 97, 106

B
Backstepping, 10, 11, 35, 40, 44, 49, 55, 64,
    114
Ball and beam, 8, 18, 27, 31, 43, 48, 64, 83,
    85
BIBS, 56
Brockett, 8, 10, 21, 22, 71
Brunovsky form, 107, 112

C
Centrifugal term, 16
CFD, 10, 35, 40, 55, 65
Chaos, 123
Chattering, 88, 119
Christoffel symbols, 16
Classification, 10, 35, 43, 55
    class I, 47
    class II, 47, 48
Complete integrability, 128
Constraint
    holonomic, 19, 27
    non-holonomic, 18, 27, 37
Controllability, 8, 17, 21–23, 40, 108, 129,
    131
Coriolis, 16

D
Degree of complexity, 36, 40
Diffeomorphism, 108, 109, 127
Distribution, 128

E
Energy, 9, 113
    kinetic, 16
    potential, 16, 28, 29
Euler–Lagrange equation, 15, 17, 19, 27,
    65
Equilibrium, 94

F
Finite time convergence, 84, 118
Forwarding, 10, 11, 44, 49
Frobenius theorem, 109, 128
Function
    Chetaev, 99
    Lipscitzian, 94
    Lyapunov, 64, 97
        common, 101
        multiples, 74, 101
        weak, 102

G
Generalized momentum, 27, 50

H
Hurwitz, 84

I
Inertial wheel pendulum, 9, 30, 65
Inverted pendulum, 8, 9, 18, 28, 42, 48, 49, 65,
    67
Involutivity, 128

J
Jacubezyk–Respondek theorem, 108

K
Kalman criterion, 129

A. Choukchou-Braham et al., *Analysis and Control of Underactuated Mechanical
Systems*, DOI 10.1007/978-3-319-02636-7,

MIX
Papier aus verantwortungsvollen Quellen
Paper from responsible sources
FSC® C105338

If you have any concerns about our products,
you can contact us on
ProductSafety@springernature.com

In case Publisher is established outside the EU,
the EU authorized representative is:
Springer Nature Customer Service Center GmbH
Europaplatz 3, 69115 Heidelberg, Germany

Printed by Libri Plureos GmbH
in Hamburg, Germany